Nuclear Waste Management in Canada

Edited by Darrin Durant and
Genevieve Fuji Johnson

Nuclear Waste Management in
Canada: Critical Issues, Critical
Perspectives

UBCPress · Vancouver · Toronto

20 19 18 17 16 15 14 13 12 11   5 4 3 2

Printed in Canada with vegetable-based inks on FSC-certified ancient-forest-free
paper (100 percent post-consumer recycled) that is processed chlorine- and acid-free.

Library and Archives Canada Cataloguing in Publication

Nuclear waste management in Canada : critical issues, critical perspectives /
edited by Darrin Durant and Genevieve Fuji Johnson.

Includes bibliographical references and index.
ISBN 978-0-7748-1708-0 (bound); ISBN 978-0-7748-1709-7 (pbk.);
ISBN 978-0-7748-1710-3 (e-book)

1. Radioactive wastes – Canada – Management – History. 2. Radioactive wastes –
Management – Government policy – Canada. 3. Radioactive wastes – Management
– Social aspects – Canada. 4. Radioactive wastes – Management – Moral and ethical
aspects – Canada. 5. Radioactive waste sites – Canada – Management. I. Durant,
Darrin II. Johnson, Genevieve Fuji

TD898.13.C3N838 2009   363.72'890971   C2009-903453-0

Canadä

UBC Press gratefully acknowledges the financial support for our publishing program
of the Government of Canada through the Book Publishing Industry Development
Program (BPIDP), and of the Canada Council for the Arts, and the British Columbia
Arts Council.

UBC Press
The University of British Columbia
2029 West Mall
Vancouver, BC V6T 1Z2
604-822-5959 / Fax: 604-822-6083
www.ubcpress.ca

To the memory of Irene Mary Kock, 1960-2001

Contents

Acknowledgments

This volume is the result of collaboration among scholars of the political, social, and ethical dimensions of nuclear waste management in Canada. Our efforts have been bolstered by the invaluable insights of several other scholars, who participated with us in a Workshop on Nuclear Waste Management held at York University, Canada, on 2-3 May 2007. From Sweden and the United States respectively, Göran Sundqvist and Patricia Fleming provided critical comments on the general direction and themes, and on specific chapters, of this volume. Lillian Trapper, in her gentle but penetrating way, also provided comments on the volume informed by her experience as an environmental policy adviser and member of the Moose Cree First Nation. Danya Al-Haydari and Julie MacArthur also provided insightful comments on individual chapters. We are very thankful for the expert insights provided by these fine scholars.

Julie MacArthur has been especially helpful throughout the process of putting this volume together. While studying full time for her PhD in political science at Simon Fraser University, she helped to organize the workshop that brought all of us together to develop this volume, provided excellent feedback on each of the chapters, formatted and copy-edited this volume, and generally took care of a million administrative tasks. Andrew Phillips, while completing his MA in political science at Simon Fraser University, also played a crucial role in our successful application to the Social Sciences and Humanities Research Council of Canada (SSHRCC) for funds to assist in workshopping and publishing this volume. And, on that note, we are very grateful to SSHRCC for its generous financial assistance with this project. This assistance has been critical in the publication of this volume and thus in the dissemination of important views that challenge the dominance of official narratives in an increasingly pressing area of public policy: nuclear energy and nuclear waste management.

Finally, we are very grateful to our families. Darrin wishes to thank his wife, Vivian, for always keeping things interesting. Genevieve is thankful, as always, for the love and support of her parents, Tsuneko Kokubo, Paul Gibbons, and Murray Johnson.

Abbreviations

AECA	Atomic Energy Control Act
AECB	Atomic Energy Control Board
AECL	Atomic Energy of Canada Limited
AFN	Assembly of First Nations
APM	Adaptive Phased Management
CEA	Canadian Electrical Association
CAMECO	Canadian Mining and Energy Company
CANDU	Canadian Deuterium-Uranium Reactor
CAP	Congress of Aboriginal Peoples
CCEER	Canadian Coalition for Ecology, Ethics, and Religion
CEAA	Canadian Environmental Assessment Agency
CEAP	Canadian Environmental Assessment Panel
CNA	Canadian Nuclear Association
CNS	Canadian Nuclear Society
CNSC	Canadian Nuclear Safety Commission
CPRN	Canadian Policy Research Networks
EIS	Environmental Impact Statement
EMR	Energy, Mines, and Resources (now NRCan)
ENL	Eldorado Nuclear Limited
FEARO	Federal Environmental Assessment Review Office
GBN	Global Business Network
H-Q	Hydro-Québec
IAEA	International Atomic Energy Agency
IICPH	International Institute of Concern for Public Health

IREE	Institute for Research on Environment and Economy
ITK	Inuit Tapiriit Kanatami
LULUs	locally unwanted land uses
NBP	New Brunswick Power
NEA	Nuclear Energy Agency
NEB	National Energy Board
NEP	National Energy Program
NFWA	Nuclear Fuel Waste Act
NGO	non-governmental organization
NIMBY	not in my backyard
NRC	National Research Council
NRCan	Natural Resources Canada
NPD	Nuclear Power Demonstration
NRU	National Research Universal Reactor
NRX	National Research Experimental
NSCA	Nuclear Safety and Control Act
NWAC	Native Women's Association of Canada
NWMO	Nuclear Waste Management Organization
OAG	Office of the Auditor General
OECD	Organization for Economic Co-operation and Development
OPA	Ontario Power Authority
OPG	Ontario Power Generation
OWMC	Ontario Waste Management Corporation
SRFN	Serpent River First Nation
SRG	Scientific Review Group
ZEEP	Zero Energy Experimental Pile

Nuclear Waste Management in Canada

1

Critical Perspectives on the Nuclear Story

Darrin Durant and Genevieve Fuji Johnson

This volume critically explores the social and ethical issues relevant to the management of high-level nuclear fuel waste and the nuclear generation of electricity. Canada and other countries utilizing nuclear energy are confronting major technical, social, and ethical issues. Commercial nuclear energy produces an ever-expanding amount of highly radioactive and chemically toxic waste in the form of used fuel bundles. The means of addressing this problem are summed up by the Nuclear Waste Management Organization (NWMO), which was mandated by the federal government in 2002 to investigate waste management options and propose a system to manage the waste: "Consistent with international standards and the regulatory regime governing management of used nuclear fuel in Canada, for the purposes of its study the NWMO has taken the position that used nuclear fuel will need to be contained and isolated from people and the environment essentially indefinitely" (NWMO 2005c, 348).

The contributors to this volume acknowledge the technical challenges of nuclear waste management but also understand the need for critical reflection on a range of important social and ethical aspects of nuclear waste management. These aspects include the process and substance of public consultation initiatives in this particular policy field; the nature and means of arriving at social and ethical frameworks within which to assess technical options, consultative practices, and decision-making processes; the meanings of recurring terms within debates about both nuclear waste management and commercial nuclear power (e.g., safety, risk, acceptability); the strategies for thinking of the long term and for proceeding in the face of uncertainty, ambiguity, and ignorance; and the significance and potential of public consultation where diverse actors, interests, and epistemologies are involved (e.g., Aboriginal peoples and public critics with long-standing engagements with nuclear issues, analysts and representatives from academia, industry, and government, and the general public).

We understand these dimensions as necessarily contextualizing the technical aspects of nuclear waste management. These dimensions provide the broader context in which this waste has been, and continues to be, created and managed. The contributors to this volume argue that this context must be examined from critical perspectives. We share healthy skepticism concerning some of the pronouncements about and treatments of the nuclear waste management issue that have, over the years, emanated from industry and government. Our volume can thus be read as a much-needed counterpoint to this dominant position, keeping in mind the disproportionate discursive power available to this coalition. Moreover, with Canada seemingly intent on building up to eight more reactors in Ontario (licensing hearings were under way as of August 2008), ostensibly as a climate-change-friendly means of increasing energy production capacity, we view this book as a timely contribution to discussions on how solutions to debates about energy politics intersect with evaluations of nuclear waste disposal proposals.

Beyond a Pragmatic Approach

Part of the need for critical perspectives on the nuclear waste management issue is that readily available government and industry accounts, although sensitive to the need to address broad concerns and present an appearance of inclusive policy making, often embody an implicitly pragmatic approach. This underlying approach was pioneered by Atomic Energy of Canada Limited (AECL), Canada's nuclear research and development corporation, and Ontario Hydro (now Ontario Power Generation [OPG]), the Ontario-based public utility that developed/built most of Canada's nuclear reactors. These organizations have, since the early 1970s (see Durant 2009a; and Durant and Stanley, this volume), repeatedly emphasized the fact that nuclear waste exists, that storage is only an interim measure, and that permanent disposal without the need for human institutional controls is the safest option. Over the years, these organizations have persistently called for the passive deep geological disposal of nuclear waste. AECL considered such an approach to disposal to be "urgent" and "necessary," adding that it also happens to be ethically fair not to burden future generations with our waste (1994a, 2-3, 336). Recently, the NWMO has also spoken in terms of the amount of waste existing, the amount projected to exist, and the necessity of dealing with the waste. Such talk forms an integral part of the nuclear industry's argument that the waste must be dealt with promptly.

Nevertheless, the NWMO uses softer language, avoiding talk of urgency and necessity. Instead, it claims that, although no "crisis" is forcing a swift decision, the "polluter pays" principle and "the sentiments and values of Canadian society" demand that this generation make provisions for managing nuclear waste (NWMO 2005c, 18). Indeed, the NWMO claims to represent the interests and sentiments of both the current generation and future generations. The NWMO accomplishes this "good-for-all-generations" position by claiming that current generations understand preparations for permanent disposal as good-faith acts to clean up our waste and to lessen the burden on future generations. Simultaneously, the NWMO's proposed Adaptive Phased Management (APM) approach to disposal leaves room for technical developments and changes in social priorities. It also makes provisions for financial resources to facilitate future choices (NWMO 2005c, 17-18).

Despite the NWMO's humility in acknowledging that only greater or lesser public confidence will be available (2005c, 17-18), and despite the elevation of broad public input and societal acceptance to an apparently decisive level of importance (Durant 2009a), we suggest that the NWMO's conceptualization of the waste problem is broadly continuous with AECL's. According to the NWMO's line of thinking, there is a requirement to manage promptly this growing inventory, in a fashion not "overly dependent upon strong institutions" (2005c, 18) and regardless of what the future may hold for commercial nuclear energy (20). Similar to how AECL and Ontario Hydro handled the waste disposal issue up until Canada's public hearings into a waste disposal concept in 1996-97, the NWMO conceptually divorces nuclear waste management from broader social and ethical aspects of energy generation itself (see Durant 2009a) while implicitly treating safety as tied up with passive disposal (see Durant and Stanley, this volume). Here we can only point to these claims as substantive points about a general continuity in approach to waste management, a continuity that spans the time when AECL and Ontario Hydro were waste disposal proponents and now covering the time when the NWMO is the waste management proponent. We suggest that the various contributions to this volume substantiate these claims of continuity, and we invite readers to consult the individual chapters and reach their own conclusions about whether these claims have been well supported.

We also note that this pragmatic "problem exists, act now" approach suffers from inevitable ambiguities, due in part to the variability in the estimates about the dimensions of the problem. The NWMO recently estimated that (as of December 2004) 1,869,163 used fuel bundles exist in storage (2005c, 350). Just how high this number will climb is subject to a number of uncertainties, including the lifetime of reactors and the number of operating reactors. In 1994, AECL projected that anywhere from 1.1 to 10 million used fuel bundles would have to be disposed of, depending on scenarios ranging from nuclear phase out to nuclear expansion (3 percent expansion in nuclear energy generating capacity after 1994 was its base case [see AECL 1994a, 114, 263]). The NWMO has been more circumspect, hoping to avoid the impression of default planning for nuclear expansion and thus using for its projections "pessimistic" estimates of the operating life of existing reactors

(2005c, 130, 141). The NWMO thus estimated either 3.6 or 3.7 million used fuel bundles (120, 350-52). The NWMO also claims that such low estimates, if compared with the 10 million used fuel bundles referred to by AECL as a base case, could be produced if changing social priorities lead to either nuclear phase out by 2012 or reactor refurbishment to an average of fifty years (223).

Nevertheless, the NWMO also entertains four scenarios (2005c Appendix 10, 393-97), three of which result in much higher waste inventories (phase out by 2012, leaving 2.5 million used fuel bundles; refurbishment of reactors to extend their average life to fifty years, leaving 4.4 million used fuel bundles by 2043; Canada's reactor sites increasing from seven to ten, and with nuclear retaining its current share of electricity generation, leaving 30 million used fuel bundles by 2200; and maintaining nuclear's share of electricity genera- tion for 200 years, but using newer reactor designs that produce less waste, leaving 15 million used fuel bundles by 2200). Although the NWMO is pol- itically astute enough not to base its estimates and projections on an expan- sion of nuclear generating capacity, it leaves little doubt that final volumes of waste are highly variable.

From the perspectives of the contributors to this volume, there is a certain arrogance in this reasoning. The AECL, NWMO, as well as other members of the dominant coalition in this policy area (e.g., Ontario Power Genera- tion, Hydro-Québec, New Brunswick Power, and long-standing private sector interest groups) tend to assert that the nuclear waste management problem can be divorced from contentious debates about whether or not to expand the current nuclear generating capacity. They draw a specious boundary around "the nuclear waste problem," defining it primarily in technical terms and treating as distinct the question of nuclear expansion. Indeed, even though the NWMO acknowledges an array of social concerns about the kinds of knowledge relevant to nuclear waste management, and about the institutions making policy decisions and the corporations generating waste, it treats such social matters as not impinging on how the technical matters are viewed or interpreted (see Durant 2006; and Johnson 2008).

Nevertheless, such specious boundaries reveal their political stripes at the level of political decision making. Despite the NWMO's claim that its recom- mendations "neither promote nor penalize" (2005c, 20) nuclear expansion, the federal government referred to its phased approach to deep geological disposal as part of "steps toward a safe, long-term plan for nuclear power in Canada" (NRCan 2007). We can now see these steps being built: as of August 2008, licensing hearings were under way for (Ontario-based) Bruce Power to build up to four new reactors at its Kincardine site and for OPG to build up to four reactors at its Darlington site near Oshawa. Beyond Canada, other nations have also used proposed solutions to the nuclear waste problem as part of arguments for further nuclear expansion (see Durant 2007a). As advertised solutions to global warming, such reactor schemes recapitulate the specious boundary drawing referred to above. Much discussion of a potential nuclear renaissance divides reactors from the broader nuclear fuel chain and thus does not seriously entertain whether mining and reprocessing activities, for instance, are carbon producing and thus factors to consider in determining whether nuclear solves the global warming problem. This volume is thus critical of both the dominant coalition's decontextualizing of the nuclear waste problem and its manipulation of what a solution to waste disposal ought to mean. All of the chapters in this volume emphasize that a wide range of important social and, more specifically, normative issues are implicated in the nuclear waste problem and how it is addressed. All of us believe that these issues must be critically explored toward the end of finding a justifiable approach to both managing this waste and deciding what the issue of nuclear waste actually means. We are committed to the political and ethical necessity of broad public deliberation. We see this volume as a contribution to what we hope will become a *public* discussion of public policy about both nuclear waste management and the possibility of nuclear expansion in Canada.

The Risk Society as a Common Theme

All of the authors in this volume engage in some way with the general theme of the importance of public deliberation in the policies of the "risk society." Ulrich Beck's (1992) thesis, that contemporary society is best understood as a risk society, provides the theoretical backdrop to this volume. Beck argued that in a risk society the distribution of "bads" (i.e., the negative consequences of industrial production) has replaced the distribution of "goods" (i.e., the fruits of industrial production) as a fundamental organizing principle. Beck pointed out that such distributive problems are so salient that society is unable to control the hazards produced by industrial production. As such, the distribution of risks becomes as much a political issue as the allocation of benefits (Beck 1996). Significantly, Beck located the causes of contemporary crises not strictly in some decontextualized notion of physical and biological effects "themselves" but also in the failures of the institutions charged with managing risk-creating activities.

Nuclear waste exemplifies the problems of a risk society not only because the very meaning and scope of what constitutes risk are at issue but also because even those with acknowledged competing conceptions of risk agree that a salient issue is the distribution of "risks." Producers and consumers of commercial nuclear energy either explicitly or implicitly seek to gain diffuse benefits by imposing costs on a localized minority, in the form of a nuclear waste disposal site. Producers thus talk of mitigating risks associated with nuclear waste management. For example, AECL outlined implementation principles for nuclear waste disposal: *safety* and *environmental* considerations ought to be primary, a community would have to *volunteer* to host a repository, decision making would have to be *shared* among relevant stake-holders, and negotiations and deliberations would have to be conducted *openly* and *fairly* (1994a, 146). For AECL, these principles not only provided a foundation on which burdens should be democratically distributed but also "should help reduce the concern over radiological risks" (256).

The NWMO also pays much attention to issues of risk distribution, hitch-ing much of its siting strategy to the idea of locating a volunteer host com-munity, allowing the relevant communities of interest to be self-defined in the negotiation process, and stressing both procedural and substantive fair-ness (2005c, 71-114, 225-35). The authors of this volume share a concern to explore different conceptions of the risks being distributed and take critical positions on dominant accounts of how risks are being distributed, how they are being conceptualized, and whether a focus on risk wrongly presumes that the public is concerned primarily with the effects of nuclear generation (i.e., technical or political) rather than with nuclear generation in the first place. Our critical perspectives are unified by concerns about conceptions of risk and processes of arriving at public policy positions based on those conceptions.

Much of the risk society literature analyzes the institutions, processes, and coalitions that create and attempt to manage risks. We retain this analytic focus because the dominant actors in nuclear decision making tend to underemphasize their role in generating risks. Moreover, where risks are brought to the front stage by dominant actors, this can be a subtle means of promoting a narrowly instrumentalist and pragmatic view of the issue in dispute. This narrow view of risks can suggest that managing them rather than generating them is in dispute. As we discuss in the following chapters, the dominant actors in Canada's nuclear field tend to skirt around the role they play in creating risks. By extension, they tend to misconstrue public values and perceptions concerning future arrangements for the management of nuclear fuel waste.

Part of our collective concern is to argue that the issue is not just the production of radioactive waste or solely how dominant actors have produced associated risks in the absence of democratic transparency and accountability. We are also concerned by the very constitution of the salient issues them-selves. Our concerns with public deliberation follow closely from these observations, for our own experience and research have convinced us not only that many public groups are capable of engaging with technical claims but also that these groups often possess distinct conceptions of what the core concerns are and ought to be. Non-governmental organizations (NGOs), for instance, have shown themselves quite adept at mounting cogent critical analyses of the scientific and engineering dimensions of waste management in a fashion continuous with the experience-based judgments of technical groups (see Durant 2007b).

Yet public groups ought not to be looked at solely as a source of addendums or clarifications to technical claims. This narrow conception of the role of public discussion often presumes that public dissent is due to either rejection or ignorance of scientific risk assessments. In contrast, we have found that public groups often reject the very way in which dominant institutional interests both frame policy questions and define what the issue is supposed to mean. The contributors to this volume thus adopt critical perspectives on the role that actors and institutions should have in the management of nuclear fuel waste. Indeed, the import of Beck's risk society thesis is that, as it becomes clear how dominant actors and institutions have failed to manage "risk" and "public issue" (and to adequately understand what they mean), the roles of expert advice and elite governance come increasingly into question. The result is not necessarily the rejection of expertise per se but what Beck calls "reflexive scientization." This is an envisioned situation in which new participants are included in decision making, pitting expert against counterexpert and creating an almost market-like environment where diverse public groups deploy expertise as a resource in highly politicized battles over factual claims and policy implications (Beck 1992, 157-63). The contributors to this volume take seriously Beck's challenge to see how the management and conceptualization of risk can be rewritten by incorporating a wider selection of voices in rendering risk decisions and understandings.

Inspired by Beck's thesis, our contributors critically explore what the role of a diversity of voices has been, and what it could or should be, in decisions about nuclear waste management. We are not necessarily unified in our underlying assumptions and prescriptions; however, we do share a commitment to broad public deliberation in decisions about both nuclear waste management and possible nuclear expansion in Canada. Yet we do not assume that even the most deliberative of forums will, in an unproblematic and unambiguous fashion, give rise to a coherent conception of the common good in nuclear waste management. At the least, the notion that a common good is the desired outcome must be subjected to scrutiny if we are to take seriously the importance of thoroughly exploring social and ethical issues. Indeed, we ask, to what extent does policy making require agreement about a common good? To what extent can policy making proceed when there are agreements to disagree or even fundamental clashes promising no resolution? We are thus skeptical about some of the ways in which the NWMO has represented public opinion in its official discourse. The NWMO, AECL, and other dominant actors and organizations have not always taken seriously the genuine conflicts among social actors. They have deployed many different strategies to incorporate, yet simultaneously disenfranchise, dissident voices. The NWMO's treatment of public input may be a perpetuation of a long-standing trend to exclude democratic decision making in nuclear energy and waste management policy (Durant 2009a).

This subtle disenfranchisement will only become more socially divisive as dominant actors move toward siting and implementing a waste management repository.

The Meaning of Social Acceptance

One question implicit in the contributions to this volume concerns precisely what social acceptance of a management approach implies. If society accepts something, then what does this acceptance imply? That policy decisions are contingent on social acceptance? That the process of negotiating what is accepted shapes policy decisions? That substantive aspects of repository design have to be accepted? That socioeconomic inducements, offered to make a particular repository design and implementation practice locally palatable, are acceptable? That good procedures ought to be followed in site selection? Or is it a combination of the above? Or something that we have overlooked?

From our perspective, it is troubling that Canada's nuclear waste management debate seems to have become a debate about the social acceptability of waste management options without much clarity about the nature and role of public acceptance in policy making. Keeping the meaning of social acceptance obscure has granted the NWMO too much discretion in its representations of the public. Given the ambiguity of the implications of public acceptance, it is vital to be critical whenever the "public will" is invoked in support of or opposition to waste management options.

We thus view with some skepticism the recent nuclear discourse in Canada. We encourage others to look behind the rhetoric of valuing public input and to ask just what kind of input is being sought and what roles that input will be allowed to play in policy making. Although each of us makes a case for public deliberation about nuclear waste management and nuclear expansion, we acknowledge that a spectrum of positions exists regarding which approaches to deliberative democracy are most effective. Deliberative democracy is paradigmatically about publicly inclusive, reasonable persuasion so that a provisionally justified compromise or altering of views can take place. But there are many ways of realizing the normative ends of deliberative democracy, and indeed those normative ends are variable. For instance, some see organized dialogue and consensus as the best means-end combination, whereas others wish to preserve space for more activist demonstration and the possibility that conflicts of interest are not resolvable. Of course, given that actual democratic practice provides space for each alternative and for mixed practices, our collective point is that their success is often contingent on the will of dominant actors and coalitions and broader contextual factors. We are concerned by the risk that dominant actors and coalitions will continue to disenfranchise legitimate dissent by seeking overzealously a consensus or a procedural purity, both of which misconstrue irreconcilable differences as differences in quantity and quality of knowledge.

All of the chapters in this volume provide evidence for the general proposition that the epistemic priority of dominant actors, coalitions, and institutions is a pervasive form of power. As Canada continues its nuclear waste management discussions, we need to remain aware of the pervasive effects of imposing a division between what is a matter of sound science and what is a matter of social acceptance (see Durant 2006). The chapters presented here show that this false dichotomy is often used surreptitiously to advance the interests of already powerful and dominant institutional actors. Whatever kind of democratic processes ensue in nuclear waste management, and the intimately connected issue of a nuclear renaissance, such a divide ought to be seen as a political coup rather than as integral to neutral assessment.

Summary of Chapters

In Chapter 2, Darrin Durant dives deeper into the trouble with nuclear energy generation and nuclear fuel waste management. He draws from the past in order to begin the critical discussion ensuing in this volume about the prospects for nuclear expansion and the likelihood of socially acceptable waste management initiatives. Thus, he lays out the political economy of the nuclear establishment as it has developed over time. In doing so, he suggests the main contours of future debates on the status of nuclear energy and the management of its waste.

In Chapter 3, Darrin Durant and Anna Stanley present a history of nuclear fuel waste management in Canada. They present both an "official" narrative and a counter-narrative concerning the development of Canadian nuclear waste management policy. The former historically dominated official channels of information dissemination; the latter exposes how the official narrative has obscured the partiality of the nuclear energy industry and falsely legitimized its goals.

Peter Timmerman goes on in Chapter 4 to discuss the history of ethics in Canadian nuclear waste debates. But he does not stop there. Indeed, he develops a notion of ethical bridges in order to better understand our obligations to future generations in nuclear waste policy.

In Chapter 5, Darrin Durant examines the Canadian Environmental Assessment Agency's environmental review of AECL's concept of deep geological disposal. He does so through the lens of Jürgen Habermas' concept of performative contradiction. Durant argues that the narrow mandate for both the environmental assessment and the NWMO's consultation process served to reinforce the pre-existing institutional influence of nuclear elites.

Genevieve Fuji Johnson continues with this line of criticism in Chapter 6, although not from a Habermasian perspective. She argues that the NWMO's consultation process involved deliberative democratic techniques that served, in fact, to advance the interests of the dominant coalition of industry representatives and government officials at the expense of important religious, environmental, and Aboriginal organizations.

Anna Stanley, in Chapter 7, examines how the nuclear energy industry, in particular the NWMO, has incorporated Aboriginal peoples into its consultation processes. She argues that the NWMO has not adequately incorporated the knowledges and experiences of Aboriginal peoples into its study of nuclear waste management options or its recommendation for a phased approach to deep geological disposal.

In Chapter 8, Brenda L. Murphy draws from Canadian and Swedish case studies an understanding of the potential future effects on Canadian communities as the NWMO begins concretely to site its Adaptive Phased Management (APM) approach. She focuses on geographically peripheral communities in Canada's north and on under-resourced NGOs as resources for critiques of the dominant nuclear establishment.

The volume closes with a chapter by Brenda L. Murphy and Richard Kuhn that situates the NWMO's APM approach in the international context. They critically assess the challenges and opportunities facing the NWMO. They argue that, despite a best practices approach, the siting process will be fraught with conflict into the foreseeable future.

Each of these contributions is critical of the dominant discourse in Canadian nuclear fuel waste management policy. We hope that this volume will encourage critical thinking more broadly about our approaches to energy generation and waste management.

2

The Trouble with Nuclear

Darrin Durant

This chapter is an opinionated guide to the nuclear establishment in Canada. It is written to provide a general background to the Canadian nuclear industry, as both a context for other chapters in this volume and a resource for domestic or international scholars, students, and concerned citizens who might not be familiar with nuclear developments in Canada. The chapter cannot be comprehensive within its bounds, but it aims to be informative. Its main goal is to articulate why so many in Canada might have trouble with the nuclear option. Understanding such misgivings is crucial to understanding the broader political-economic context within which nuclear waste management discussions are carried out.

Inevitably, the chapter will be read as advocating an anti-nuclear stance, so let me be clear about what kinds of commitment inform the chapter. I am sympathetic to the claim that the social return on investment offered by nuclear power has neither matched nor will match the predictions of nuclear proponents. Yet the prime normative commitment informing this chapter is dissatisfaction with the standard of democratic decision making that has accompanied nuclear power. Unfortunately, this standard has carried over, albeit in a reinvented form, to the nuclear waste management issue. Energy options are social choices, and policy making about nuclear projects in Canada has fallen well short of the kind of democratic standard that ought to prevail in regard to such important choices. Moreover, this democratic deficit means that many concerns of critics, such as health, safety, and environmental concerns, have received inadequate attention. If I am advocating anything here, then it is that, where an interrogation of policy making for its democratic credentials reveals serious inadequacies, it is the job of social science to make such inadequacies plain. Rather than offer simplistic policy recommendations, I connect identified inadequacies with the nuclear option to the operation of power relations. What emerges is a history of little democratic accountability, one that suggests the lack of wisdom in leaving decisions about nuclear projects in the hands of nuclear industry actors and their bureaucratic supporters.

This chapter draws heavily on Canadian critics of nuclear power and their claims about a Canadian "nuclear establishment." Mehta thus describes Canadian nuclear decision making as "even more concentrated, bureaucratic and inaccessible than most" (2005, 43). Mehta endorses Jacob's definition of a nuclear establishment as "a relatively stable set of relations among members of groups and institutions that promoted and benefited from the development of nuclear power and other nuclear technologies" (1990, 4). The concept of a "nuclear establishment" shares a family resemblance with similar terms, such as "subgovernment," "subsystem," and "iron triangle" (McCool 1998, 551-52). Such terms have been standard parts of critical positions on nuclear power since at least the late 1960s, both in Canada (Babin 1985) and internationally (Surry and Huggett 1976; Welsh 2000). They form part of a neopluralist political critique, which gained prominence in the early 1960s and which posits multiple elites ruling particular policy areas as separate mini-elites (Mawhinney 2001, 200-2). Careful analytical use of such notions means inter-relating exogenous sociopolitical conditions over time to the strategic actions that mini-elites engage in to control conflict. The normative implication of "establishment" metaphors is that mini-elites strategically minimize public involvement in decision making.

This chapter utilizes a periodization of Canadian nuclear power history. In the period 1945-57, the legal and institutional structure of the nuclear industry was established. Research on prototype reactors was undertaken, and immense political stability characterized the nuclear scene (due in part to the 1935-57 reign of the federal Liberal government). The period 1958-93 saw the rise and decline of the nuclear option. Prototype and commercial Canadian Deuterium Uranium (CANDU) reactors were constructed, with Canada's peak reactor start-up point occurring in the mid-1980s. Nevertheless, there were no new domestic reactor orders after 1974 and only one foreign reactor sale (to Romania in 1978) in the period 1977-89. The third period encompasses 1994-2008 and centres on electricity market restructuring, nuclear waste disposal and climate change issues, and a renaissance of claims that nuclear expansion is (again) at hand.

C.D. Howe and the Nuclear Option, 1945-57

C.D. Howe (Liberal cabinet minister, 1935-57) has been described as the "architect of modern industrial Canada" (Babin 1985, 41) and as having had "mastery of the atomic field in Canada" (Bothwell 1984, 178). During World War II, Howe oversaw the establishment of dozens of crown corporations. In Canada, crown corporations have been relatively coercive but otherwise typical policy instruments for maintaining monopoly situations and reconciling conflicting policy objectives (Doern and Wilson 1974; Pritchard and Trebilcock 1983). By the postwar period, Canada's electrical utilities industry was dominated by monopolistic enterprises, making it difficult for "civic populists" to exert democratic control (Armstrong and Nelles 1986). Howe used crown corporations to establish Canada's nuclear projects.

Although nuclear developments *continued* this tradition, they also exacerbated discontent with such governance style, in part because nuclear developments were shrouded in secrecy and security as a spin-off from the Manhattan Project (Bothwell 1984, Chapter 5; Eggleston 1965). Concerns have also lingered that commercial nuclear power can never be unambiguously separated from its military origins. In Canada, though, links between civilian reactors and nuclear weapons gained prominence only after India's 1974 atomic bomb test (Babin 1985, 145-47; Knelman 1976, 164; Mehta 2005, 38-41). India's atomic program had been enabled by the transfer from Atomic Energy of Canada Limited (AECL) of a Canadian research reactor to India in 1956 (from which plutonium had been produced). Subsequently, AECL supplied both heavy water and technical assistance to the Indian nuclear program. Canada's unintentional role in the bomb test caused a domestic political crisis, fuelled by public disquiet about the possibility of diverting spent fuel waste to military uses and the efficacy of non-proliferation policies (Bratt 2006, 87-97, 117-28).

Mackenzie King's Liberal government practised Manhattan Project-style secrecy from an early point. To control uranium prospecting, extraction, and production in Canada, the government bought up shares in Eldorado Gold Mines in 1942. The company became the first nuclear crown corporation in January 1944 and was renamed Eldorado Nuclear Limited (ENL, now known as Canadian Mining and Energy Company [CAMECO]) (Bothwell 1984). When Howe summarily announced to the House of Commons in December 1945 that Canada was not interested in atomic bombs, he showed how nuclear policy was already the province of very few (Bothwell 1984, 169). Yet the kind of centralized government control that would come to define nuclear policy was not unusual in the postwar period. Certainly, Howe was no stranger to crown corporations. Hence, as draft legislation to control the atom circulated in August 1945, it was taken for granted that only Howe and the prime minister would make atomic policy.

Howe continued to exert control until he lost office in 1957 (Bothwell 1984, 168-97). The Atomic Energy Control Act (AECA) of October 1946 utilized the federal prerogative to assume control over a resource area where it is deemed in the national interest. The Atomic Energy Control Board (AECB, now the Canadian Nuclear Safety Commission [CNSC]) was established as both regulator and developer of nuclear power in Canada (Doern 1977). This conflicting mandate was embodied in Howe's hand-chosen first

AECB president, C.J. MacKenzie, then acting president of the National Research Council (NRC), which was developing nuclear projects at Chalk River near Ottawa.

Having originally told the House of Commons in early 1946 that the AECB would be a policy board, Howe nevertheless reassured W.J. Bennett, who was running ENL as a profit-making venture, not to be concerned about the apparently sweeping powers of the AECB. The AECB was to be limited to technical functions, unable to exercise its powers without approval of the NRC and ultimately Howe (Bothwell 1984, 172, 181-83, 240). The AECB was "largely symbolic" and became a "ratifying authority for decisions taken elsewhere" (Bothwell 1988, 77). Part of the problem for the AECB in its early life was that it was reliant on its clients, the nuclear industry, for staff and was chronically understaffed (Sims 1981, 60-70). Critical analysis of the AECB has thus routinely concluded that it was unable to act as an "honest arbitrator" (Johannson and Thomas 1981, 442), resulting in an industry that basically regulated itself, with the AECB "simply granting the industry its institutionalized blessing" (Babin 1985, 59). Moreover, critics claim that the AECB developed a pro-industry "siege mentality" (Mehta 2005, 49) in response to conflict.

AECL was created in 1952 to address the AECB's conflicting mandate. AECL took over the NRC's basic research operations at Chalk River, first as a holding company reporting directly to the AECB (and thus to Howe), but then in 1954 as a wholly separate crown corporation (via amendments to the AECA), reporting directly to Howe. Bennett, who had passed through ENL and the AECB, became AECL's first president. Interlocking AECB/AECL board members remained common into the 1970s (Sims 1981, 24-34, 60-69). AECL was given the role of promoting and developing nuclear power, initially via the domestic reactor market (Bothwell 1988). Over time, AECL shifted toward a focus on international reactor sales as well (Bratt 2006; Doern, Dorman, and Morrison 2001b; Finch 1986; Sierra Club of Canada 2001, 28-48).

Howe saw the 1954 AECA amendments as expanding the promotional aspect of nuclear power operations at the expense of the regulatory aspects, which he considered potentially unnecessary (Sims 1981, 42). For Howe, as for those who had secretly worked on the experimental reactors at Chalk River (the Zero Energy Experimental Pile [ZEEP] went critical in September 1945, the Nuclear Research Experimental [NRX] reactor in July 1947; the future National Research Universal ([NRU] reactor of 1957 had received approval and was being worked on by December 1950), AECL was a vehicle for pursuing "power piles." W.B. Lewis, whom later AECL president J. Lorne Gray (1958-74) described (1987) as the most significant contributor to nuclear power development in Canada, had already proposed to the NRC (as early as 1946) that commercial nuclear power was possible, would be cheaper than coal, and had the distinct advantage over hydroelectric that it could be made available anywhere (Bothwell 1988, 172-76, 182-89). Howe and Bennett had initiated secret discussions in early 1952 with the Ontario premier (Leslie Frost, Conservative) and the provincial utility, Ontario Hydro (now Ontario Power Generation [OPG]), about co-operation between the future AECL and the province regarding a nuclear-powered electrical generating system for the province (Bothwell 1988, 189-208; Freeman 1996; McKay 1983, 34-52; Swift and Stewart 2004).

Despite the international rhetoric of "peaceful atoms" since the late 1940s (Welsh 2000), domestic policy focused on profitable atoms. "Nation-accentuating" motivations are typically cited to explain the attractiveness of the nuclear option: national independence and energy diversification, the prestige of developing a high-tech industry, the lure of cheap electricity due to an abundance of uranium, and competition with the United States and United Kingdom. Justifications for the centralized institutional governance of atoms typically focus on security, secrecy, the complexity and danger of the technology itself, and the prevention of weapons proliferation (Sims 1981, 23, 179). For Howe, the centralized hierarchy was justified by the need for a unified top command (Bothwell 1988, 69-82, 144). Gray recalled that having an interlocking administrative arrangement that reported to the one place simply meant that "it was possible to operate with little, if any, friction" (1987). Yet for many critics, that lack of friction implied little democratic accountability in regard to nuclear decisions (Babin 1985, 23-24, 45; Knelman 1976, 7-16; Mehta 2005, 45).

Specific Canadian traditions also influenced developments, in the form of parallel (and overlapping) traditions of mini-elite decision structures and the favouring of megaprojects. Howe, for instance, was a typical figure in what Pross (1985) has argued was the age of the "mandarins" from 1935 to 1965. The role of public discussion and Parliament in the policy process was at a low point in this period. Single-party dominance stifled both inter- and intraparty debate, and only those pressure groups that both retained close connections to policy structures and avoided appealing to broader constituencies were successful (Pross 1985, 244-46). For the record, the federal Liberals were in power from 1935 to 1957, and in Ontario (which would come to have twenty of Canada's twenty-two reactors), the Progressive Conservatives were in power from 1943 to 1985.

Technical megaprojects were also familiar. Armstrong and Nelles (1986) showed that, from 1830 to 1930, the rise of giant utility monopolies had confronted Canadians with distinct political and economic challenges. Various public-private organizations were created to regulate these monopolies, ranging from outright government ownership to weak oversight. Armstrong and Nelles argued that regulations served to legitimate the monopolistic situations. Canadian economic historians, as early as 1956, noted the preponderance of public enterprises in Canada (Musolf 1956, 421). The mandarins and the megaprojects fused in the postwar nuclear undertaking, influenced by the legacy of the Manhattan Project. Yet this fusion was also influenced by the resource politics surrounding hydroelectricity and coal in central Canada (Ontario and Quebec) at the time.

Howe's efforts from 1944 to 1947 to develop nuclear power as a possible energy source occurred prior to the discovery of major oil and gas reserves in Alberta in 1947. The immediate context for Howe was coal dependency in central Canada (he initiated a Royal Commission on Coal in October 1944, chaired by W.F. Carroll). Dales (1953) showed that in the 1940s and 1950s central Canada depended on the importation of coal and petroleum for its fuels (e.g., used for industrial heating), but otherwise most of its industrial motive power needs were met by hydroelectricity (and central Canada supplied 75 percent of Canada's hydroelectric power). Federal and provincial governments prevented exporting power to the United States to encourage the movement of industry to central Canada, and the federal government was always concerned to keep electricity and fuel prices uniform in Ontario and Quebec.

Hydroelectric was the trickier problem. Armstrong (1981) showed that into the 1940s the federal government had been unable to resolve issues of constitutional jurisdiction over hydroelectric power, leading to federal-provincial political squabbles. Although there had been public ownership of Ontario Hydro since 1904, in Quebec there was a mix of the publicly owned Hydro-Québec since 1944 and private utilities, which would not be nationalized until the 1960s (Bothwell 1988, 332-33; Doern and Gattinger 2003, 27-28). With both federal and provincial governments using the utilities for economic development purposes, issues of unity, equity, and ownership created resource politics. Indeed, hydroelectric power would decline from an estimated 90 percent of Canada's installed electrical capacity between 1920 and 1950 to 80 percent in 1960, 60 percent by the mid-1970s, after which it has held steady at roughly 60 percent (Canadian Electrical Association 2006, 5; EMR and CEA 1992).

Nuclear reactors contributed to this decline, with their "portability" offering a means for the federal government to intervene in federal-provincial jurisdictional squabbles over hydroelectricity and dependency on coal imports. Notably, coal-fired thermal generation as a share of Canada's electricity mix did not rise above 7 percent until the 1960s (Coal Industry Advisory Board 1995, 49). In Ontario, it was not until 1951 that "the first of six coal-fired generating stations locked into the grid, marking an end to the unchallenged reign of hydro-electric energy" (McKay 1983, 33). McKay (1983, 320-52) notes that the challenge to hydroelectric went even further. By the early 1950s, Ontario's residential and commercial sectors were rapidly expanding, and it was thought that hydroelectric generation might not meet the demand. Cheap western Canadian natural gas, arriving in the mid-1950s, threatened the water heating and cooking ranges market controlled by hydroelectric power. Ontario Hydro thus launched an aggressive marketing campaign to shift user patterns toward electricity-dominant activities and practices. This campaign partially fuelled a 600 percent increase in electricity consumption between 1957 and 1974.

Electricity at the time was seen as a natural monopoly (Jaccard 1995), thus involving significant economies of scale and pressures for public ownership (Doern and Gattinger 2003, 28). Nuclear reactors offered a new and distinctive tool that could facilitate what was already a standard type of government action: intervention in the name of provincial economic development.

The Rise and Fall of the Nuclear Option, 1958-93
Howe departed from political office in 1957, but he had done much to place the nuclear project on a strong institutional footing. Construction of two prototypical reactors commenced in 1958 (Nuclear Power Demonstration [NPD], completed in 1962) and 1960 (Douglas Point, completed in 1968). In 1959-60, what would become the largest nuclear industry lobby group in Canada, the Canadian Nuclear Association (CNA), was formed. Ironically, Gray, then head of AECL, refused to join on the ground that an identifiable nuclear industry neither existed then nor would in the future (Weller 1990, 9). Although federal energy policy has always focused more on oil and gas (Doern and Gattinger 2003, 27-31), Gray turned out to be wrong. Ontario Hydro's commercial nuclear power expansion program drove the nuclear option in Canada, commencing with the construction of Pickering 1 in June 1966 (the first operating commercial reactor, coming online in June 1971) and finishing with Darlington 4 in June 1993 (Sierra Club of Canada 2001, 28-29).

Ontario Hydro eventually built twenty reactors. Hydro-Québec and New Brunswick Power (NBP) now (as of 2008) operate one reactor each. These utilities derived their influence from their monopoly of generation and transmission facilities, their exemption from federal income tax, and their provincially guaranteed loans. Compared with private enterprises, public utilities could borrow cheaply, increasing their ability to compete for scarce capital and engage in risky and low-return projects (Jenkins 1985). The public utilities functioned as arms of province-growing efforts and benefited from the rationale that electricity is both a warranted natural monopoly and a public good (Jaccard 1995). Yet Ontario Hydro became an exemplar of the notion that the already powerful use their institutionalized power to perpetuate their own preferences.

An early history of Ontario Hydro predicted that its future would depend on nuclear power and championed its success (to 1960) as predicated on its commitment to "the people" (Denison 1960, 274, 277). Later accounts attribute the decline of Ontario Hydro to nuclear overexpansion and its independence from the political process – that in fact it went "out of control" (Solomon 1984). Yet Ontario Hydro was not unique in pursuing an ambitious nuclear expansion program. Similar programs also developed in Western Europe (Romerio 1998; Soderholm 1998) and the United States (Joppke 1992-93).

Critics of Canada's program of nuclear expansion have claimed that the expansion in electricity supply itself was ill conceived. McKay (1983, 36-58) notes that Ontario Hydro increased its generating capacity fourfold between 1960 and 1975. Rather than holding an idealized notion of independent public demand for electricity driving supply, Ontario Hydro itself contributed to creating demand (e.g., by running electricity campaigns, entering the electrical space heating market, and neglecting changing consumption patterns and capital availability). Most commentators agree that Ontario Hydro's load forecasters, who in the 1960s predicted a 7 percent growth rate in electrical demand to the year 2000, got the forecast wrong. Moreover, critics claim that Ontario Hydro failed to relinquish the commitment to satisfying this projected large growth until the late 1980s. Critics thus conclude that Ontario Hydro overinvested in nuclear capacity to meet over-inflated power needs (Daniels and Trebilcock 1996, 6; Doern and Gattinger 2003, 31-37; Jaccard 1995, 586; Martin 2000, 23-24; McKay 1983, 208; Solomon 1984, 74-78).

Democratic scrutiny was also lacking. McKay (1983, 207-20) notes that Ontario Hydro, even as it pursued more economically and environmentally costly nuclear capacity, was aware of the undeveloped and cheaper (to build and operate) hydroelectric capacity available. Solomon (1984) argues that Ontario Hydro acted as an unregulated monopoly, thwarting any attempted governance oversight. Even sympathetic analysts regard Ontario Hydro as having escaped the public transparency requirements of democratic regulation (Daniels and Trebilcock 1996, 6). By the 1970s, attributions of a democratic deficit were prominent features of international opposition to nuclear power (Surrey and Huggett 1976, 287). Lovins' much-cited critique of nuclear power, and the general "hard" energy path, attributed many ills to the "far-away, bureaucratized, technical elite" who formed an "incestuous" government-industry constituency (1976, 92-93). Canadian accounts also criticized democratic deficits and bureaucratic elitism. Knelman (1976) thought that the nuclear problem involved a technocratic ideology, biased experts, and lack of information access. Babin (1985) thought that the nuclear problem involved technocratic corporations managing vast information and production systems and using the umbrella myth of complex nuclear technology in the service of the public good to depoliticize the issue. This motif of a democratic deficit exists as a constant feature in diverse critiques and is often cited as a factor compounding more specific complaints.

The high degree of centralization regarding nuclear policy making in Canada partly accounts for these themes. The federal government retains almost complete "jurisdiction over nuclear matters" (Doern, Dorman, and Morrison 2001a, 4). While Senate and House committees have engaged in discussions about nuclear matters, particular ministries have solidified the role of government at the centre of nuclear policy. Since 1993, the central organ has been Natural Resources Canada (NRCan). There has been minimal variation in the locale of federal responsibility: the Privy Council Committee for the Ministry of Science and Industry (1946-65), the Department of Mines and Technical Surveys (1965), and the Department of Energy, Mines, and Resources (EMR, 1966-93). EMR was renamed NRCan in 1993. Given bureaucratic centralization, much critical attention has been directed toward the regulatory structure.

Doern's early characterization of Canadian nuclear regulation as professionally open (high levels of trust existed between networked elites) but democratically closed (little provision for public participation) has been seminal in critical scholarship (1977, 33). Attempts in 1977 to rewrite the AECA to allow for greater public involvement failed (Johannson and Thomas 1981), leading Doern to predict correctly an erosion of the credibility of the regulatory structure (Doern and Morrison 1980, 55; Doern 1981, 68, 196). The central concern was always the cosy relationship between industry and the AECB. Three broad means of bringing into doubt regulatory independence have been common in the Canadian literature (typically, all three are used in critiques).

Doern's early work is the archetype for a procedural critique, focusing on transparency and accountability in the regulatory process as required for credibility to exist. A second critique emphasizes the problem of cross-cutting government commitments, such as commercial and regulatory activities. Johannson and Thomas (1981) discuss Canada's participation in a uranium cartel as an example of the executive arm of government using nuclear regulatory powers to advance commercial interests, but they also suggest that the more subtle problems of commitment to nuclear non-proliferation, and foreign reactor sales, sit uneasily with heavy regulation. Others have pointed to environmental assessment hearings into reactor licensing at Point Lepreau in 1975 (Salter and Slaco 1981) and Pickering in 1992-94 (Mehta 2005, Chapters 5 and 6) as being more about symbolic rituals and public education than public participation, suggesting that governments use the regulatory process to surreptitiously legitimate their electricity production preferences.

A third critique focuses on the pragmatic difficulties that regulators face in extracting themselves from the regulated industry. Canada's early nuclear developments witnessed a relatively small community of experts circulating between jobs with the AECB, AECL, and relevant government ministries.

Personal networks thus blurred promotion-regulation lines. A common refrain has been that a similarly technocratic outlook, granting technical experts wide discretion to define what an issue means and how to make assessments, was shared among all nuclear sector workers (Babin 1985; Mehta 2005; Schrecker 1987). Recent commentary draws on such past assessments to create a sense of bureaucratic inertia. Martin claims that a lack of public hearings continues a long tradition of "accommodating the wishes of the nuclear industry" (2003, 36) and that the nuclear establishment "has refused to allow any public consultation" about nuclear development (2000, 4).

As Canada's nuclear expansion proceeded, the gap between industry development and public accountability crystallized as a number of contentious issues. One concerned whether nuclear power is an economic energy production choice, given the extent of federal subsidies for nuclear projects. Although intervention in energy sectors has been a common but contentious aspect of Canadian industrialization since federation (Armstrong and Nelles 1986; Doern and Gattinger 2003; McDougall 1982), one of the crucial periods for commercial nuclear power coincided with federal policies emphasizing centralized rule and subsidization (Durant 2009b). Fourteen commercial reactors came online in the period 1971-84, and in this period the federal Liberals under Pierre Trudeau (1968-79 and 1980-84) embraced centralization, acting as an entrepreneur via pricing policies and subsidization of megaprojects. The Liberals' ill-fated 1980 National Energy Program (NEP) would eventually push this trend too far, with the Brian Mulroney Progressive Conservatives coming to power in 1984 on the backs of decentralization and deregulation policies (Duquette 1995). Nevertheless, the nine reactors that came online during the Mulroney period (1984-93) had all been approved during the Liberal reign.

The debate about the economics of nuclear power is infinitely complicated because a host of assumptions are built into calculating "the streams of income and expenditure at different times" in the life cycles of nuclear projects (Thomas 2005, 26-27). This life cycle stretches from uranium mining, through reactor research, development, construction, and operation, and ultimately to decommission and waste disposal. Critiques of the privileged relationship among governments and the utilities have directly informed negative assessments of the economics of nuclear power. Jenkins (1985) argued that supporting electrical monopolies encouraged overconsumption of electricity, led to electricity rates below the cost of production, facilitated unwise investment in capital-intensive technology such as nuclear reactors, and created substantial economic losses (he estimated as much as 1 percent of Canada's gross domestic product, or more than a billion dollars, by 1981). In critiquing utility monopolies, Jenkins recommended raising electricity costs "so as to recover the full economic opportunity costs of the resources used" (1985, 497). Later advocates of electricity deregulation and privatization would also claim that "artificially suppressed prices" (Trebilcock and Hrab 2005, 140) hinder energy reform.

Ontario has always been the most prominent point of reference for debates about economic merit. McKay (1983) argued that Ontario Hydro's share of provincial government capital spending increased from 17 percent to 70 percent between the mid-1960s and the late 1970s, an increase that he connected to severe cuts in normal spending on "schools, hospitals, roads and public buildings" (168-69). By 1975, the provincial treasury borrowings from public bond markets had ballooned to $1.86 billion, 80 percent of which was borrowed in Ontario Hydro's name (163). The consequence of Ontario Hydro's "unreality syndrome" was the "distortion of economic and social priorities" (175). Solomon (1984, Chapters 4 and 5) also critiqued the Ontario Hydro expansion program, claiming that investing $12 billion for the Darlington nuclear plant created 300 jobs but that the same investment would have created 57,000 jobs via one Bell Canada or 176,000 jobs via four GM Canada plants (49). Indeed, every job created in the nuclear-dominated electricity sector "eliminates six or seven in other areas" (51). The growth of the Ontario Hydro bureaucracy itself was also considered out of alignment with other economic sectors, with accusations of inflated salaries and a bloated bureaucracy (63-67).

To the extent that democratic accountability is about transparent influence, critics argue that the lack of transparency remains a prime reason that the nuclear option was not abandoned in its infancy. Schrecker quoted Norman Aspin, president of the CNA in 1980, describing the private sector as being in a "unique position" to talk to politicians, with Ontario Hydro and AECL as two of its biggest customers (1987, 28n71). Indeed, Schrecker was entirely representative of Canadian critics when he argued that a combination of government commitment to nuclear power, and privileged access of AECL to bureaucracy, accounted for the initial political clout of the nuclear industry. But over time, argued Schrecker, mutually supportive relations between key actors, and a strong sense of identity among those actors, account for the continued influence of industry groups (11-12). Schrecker is worth quoting at length on this point: "The nuclear industry's 'political clout' initially stemmed from the strong commitment of senior members of the government and the unrestricted access to key officials enjoyed by AECL. However, the real economic and political strength of the industry over the long term arises from the network of private suppliers and the role of Ontario Hydro as a major capital investor ... [There exists] a strong perception of identity of interest among government, these public enterprises and the many private firms in the field ... This cohesiveness is reinforced through the Canadian Nuclear Association" (11-12).

The fall of nuclear power, claim many critics, stems from economic and political realities catching up to the nuclear option. Hence, in 1975-76, just as the CNA was forecasting 180-213 CANDU units by 2000 (1975, 1, 28), the Ontario provincial government responded to a poor credit review (due mostly to Ontario Hydro's capital borrowing and planned expenditure on nuclear projects) by forcing Ontario Hydro to trim by 13 percent its projected ten-year, $40 billion capital expenditures program. Ontario Hydro's response was to stretch out its spending, introducing delays in reactor construction programs but preserving the faith in nuclear expansion to meet electricity demand. Yet by 1977, the average growth in electricity demand had begun to fall from the late-1960s level of 7-8 percent. By the late 1970s, projections varied from 2 to 4.5 percent growth by the year 2000 (McKay 1983, 164-71). Indeed, Thomas (2005, 12-17) notes that nuclear power is one of the most capital-intensive energy forms, with about two-thirds of generating costs in fixed costs and the remaining third in running costs. Thomas also notes that cost forecasting for nuclear projects has been historically unreliable. In Canada, the paradigmatic instance is the Darlington generating station. Construction of the four 881 Megawatt (net) reactors commenced in 1978, was meant to be completed by 1983, but was not completed until 1993. Cost estimates inflated from an initial $2.5 billion to an eventual $14.4 billion (Adams 2000; Martin 2003, 16-17; Pembina Institute 2004, 108-13; Trebilcock and Hrab 2005, 124). When Darlington finally came online in the early 1990s, accompanied by 30 percent rate increases in electricity, its capacity was already surplus to Ontario's needs (Doern and Gattinger 2003, 36).

Concerns about reactor safety itself have also intersected with critiques of decision-making integrity. Knelman's (1976) characterization of nuclear technology as "unforgiving" was meant to point to whether or not a hazardous piece of technology was being managed appropriately. Even independent analysts concluded that major shortcomings existed in Canadian reactor performance studies that limited their usefulness for reaching safety assessments (Oxman et al. 1989, 50). Significantly, the homogeneity of Canada's nuclear industry was cited as responsible for a lack of independent regulatory oversight of reactor safety (Oxman et al. 1989, 51). Indeed, as Arai (2001) noted, when an inquiry was held into Ontario Hydro's 1997 decision to lay up seven of its twenty reactors, critics presumed that the safety case against CANDUs had already been made. They thus concentrated on mismanagement, a consequence of little democratic oversight, as the cause of the organization's problems.

Concerns about the safety of nuclear projects have not been limited to reactors. Environmental non-governmental organizations (NGOs) have been prominent in focusing on the "waste generation, atmospheric releases, impacts on water quality and water use, and landscape and ecosystem impacts of nuclear energy production" (Pembina Institute 2006, 3). The claim is that nuclear power is not a "clean" energy source, for each stage of nuclear energy production is associated with severe environmental impacts and risks. Although this volume concentrates on the waste products of nuclear reactors, no less important are the impacts and risks associated with uranium mining, milling, and refining, conversion and fuel fabrication, and nuclear power plant operation itself. These impacts and risks include the creation of environmental hazards, occupational health and safety issues, and chronic exposure to low-level radiation by power plant workers and communities in the vicinity of nuclear projects. Moreover, some communities have been far more affected than others, such as Aboriginal peoples (see Stanley, this volume).

Consistent with the argument of this chapter, such concerns intersect in important ways with issues of democratic accountability and transparency. The Pembina Institute's study of environmental impacts notes that it most likely underestimated the severity of impacts and hazards associated with nuclear projects because of "significant gaps in publicly available information" (Pembina Institute 2006, 3). Similarly, others claim that estimates and assessments of health risks to power plant workers and communities near power plants, especially women, systematically underestimate or simply do not explore the effects of low-level radiation (Beal et al. 1987). The cause of this situation was the fact that Ontario Hydro in particular had not been "brought under democratic control" and that "in many decisions public review is absent" (136, 78).

Nuclear projects have thus been subject to a host of critiques, focusing on the economic, environmental, safety, and regulatory dimensions of nuclear power. All four critiques intersect with the issue of democratic accountability and transparency. It is thus no coincidence that nuclear power stagnated, from at least the late 1970s on, as direct intervention in energy policy began to lose its political tractability. Less centralized rule regarding energy policy and less inclination to intervene in industrial development via subsidies and overt state protection, at least as applied to the domestic reactor market, meant that proponents of nuclear power found it much more difficult to avoid the demands of democratic accountability (Durant 2009a).

In Canada, these trends followed a period in which the oil crisis of 1973 and 1979-80 had made energy security of paramount importance in government planning. The Trudeau Liberals had favoured centralized, interventionist policies that used subsidies to support megaprojects. Yet the 1980 NEP was attacked as going too far toward federal control and monitoring of private and provincial interests (Desveaux, Lindquist, and Toner 1994, 502-5, 512-14; Doern and Gattinger 2003, 21-39; Duquette 1995, 237-41). Mulroney thus came to power in 1984 not only with a policy to abolish the NEP but also with a broader policy platform of tempering federal-provincial conflicts and respecting provincial jurisdictions (Duquette 1995, 239-40,

248-49). Indeed, Doern and Gattinger (2003, 35) note that the NEP strained federal-Ontario partnerships well into the 1990s since it left a bitter taste of how not to implement energy policy. As Bratt (2006, 169-73) noted, the Mulroney government would reduce its commitment to nuclear subsidies in the 1985-90 period as part of an overall non-interventionist energy policy approach.

With close institutional and informal relations between federal and Ontario authorities having fostered initial nuclear projects, the mid-1980s shift in policy attitude toward non-interventionism suggests that some of the building blocks of the nuclear project had been undermined. Or they were no longer strong enough to override uncertainties and troubles elsewhere with the nuclear project. The typical rationale for electrical monopolies, that problems of scale made them natural, was also in decline among industrialized nations at this time. Deregulation trends centred on claims about the inefficiencies attending government direction of investment and its general economic management capacities. These arguments seemed pertinent in the case of the overinvestment and surplus capacity problems with the nuclear option in Canada (Jaccard 1995, 581-86).

A suggestive analogy is that, as the balance in the broader political scene began to favour devolved political direction of the energy sector, the strength of the nuclear option in Canada began to match the pattern revealed by Soderholm's analysis (1998, 216-17). Opposition to nuclear projects has been more successful in those countries with substantial devolutions of authority and/or weak central government control. This analogy is most suggestive when applied to Ontario. Although the Progressive Conservatives ruled Ontario from 1943 to 1985, from 1975 to 1981 they formed a minority government. An alliance between the Liberals and New Democrats created the ground for extensive contacts between environmental groups and the legislative opposition. Questions of resource development, environmental regulation, and the clientele relationships between legislative bodies and waste producers received increased scrutiny during this period (Lyon 1984). The Liberal minority (1985-87) and majority (1987-90) governments would also see internal divisions, over how to balance the goals of environmental protection and economic growth, curtail Ontario's once rabidly pro-nuclear energy policy (Winfield 1994, 134-41). The New Democratic Party government (1990-95) would eventually bring some of this environmental activism to fruition by imposing (in 1990) a moratorium on new nuclear reactor development and cancelling Ontario Hydro's twenty-five-year plan to build more reactors (Swift and Stewart 2004; Winfield 1994).

As the 1990s began, the nuclear option seemed to be in retreat. Maurice Strong, the incoming CEO of Ontario Hydro in late 1992, supported the New Democrats' moratorium on nuclear power. Strong noted that nuclear megaprojects had created Ontario Hydro's financial difficulties because their performance had not lived up to expectations, and he considered privatization of Ontario Hydro an option (Standing Committee on Government Agencies 1992). By 1993, a distinct rationale for such privatization was being discussed: the Ontario government could separate itself from the $34 billion in debt accrued through over-expanding the nuclear option. Yet doubts were already being expressed about whether there would be a market for the nuclear assets of the utilities (McMurdy 1993).

Liberalized Markets, Nuclear Waste, and Green Reactors, 1994-2008
The failure of the nuclear monopoly was evident in the 1990s as 35 percent of Ontario Hydro's electrical revenue was being put toward debt repayment (nuclear overexpansion and cost over-runs). The retail price of electricity had been frozen by the Ontario government in 1993, and this freeze remained in place until rates were opened to competition in May 2002. When the Progressive Conservatives were elected to Ontario office in 1995, they set in motion plans to restructure the Ontario electricity market. Such moves had already been undertaken in California and the United Kingdom most prominently (Doern and Gattinger 2003). Ontario Hydro was split (in April 1999) into OPG, with ownership of generation assets, and Hydro One, with ownership of the transmission grid.

Yet electricity market restructuring, which initially prompted suggestions that monopoly's moment might have passed (Daniels 1996), ultimately failed (Iacobucci, Trebilcock, and Winter 2006, 29-48; Pembina Institute 2004, 1-4; Swift and Stewart 2005; Thomas 2004, 5-6; Trebilcock and Hrab 2005). Prior to the market being opened to wholesale and retail price competition on 1 May 2002, the potential monopoly power of OPG was curbed by legislative mandates that OPG would pay a rebate to consumers where its wholesale price exceeded 3.8 cents per Kilowatt hour (kWh). OPG also had to divest itself of a significant portion of its generating assets in order to reduce its market share to 65 percent by 2004 and 35 percent by 2010. OPG thus leased the Bruce A and B generating stations to a consortium in 2000, which originally had British Energy with over 80 percent shares and CAMECO with 15 percent. Yet when British Energy went bankrupt, CAMECO increased its stake in 2004 to more than 30 percent. CAMECO is an amalgamation of the crown corporation ENL, which had been privatized in 1988 and amalgamated with the Saskatchewan Mining Development Corporation. CAMECO thus consumes its own product (uranium).

When the electricity market did open in May 2002, it remained open only until November 2002. Soaring electricity prices led to a freeze on retail rates, while, in the same month that the market opened, the courts blocked the privatization of Hydro One. Howard Hampton (2003), until March 2009 the leader of the New Democratic Party (NDP) in Ontario, argued that the history of electricity privatization and deregulation (running through Chile, the United Kingdom, the United States, and now Canada) *inevitably* witnessed such problems. Others noted that the short-term "market power" of OPG was not changed by restructuring (Dewees 2001, 169). Some argue that this is the root of the trouble. Thus, those in favour of privatization, and opposed to monopoly, have claimed that the failed restructuring means costs will be filtered through to general taxpayers, via subsidized electricity prices and government debt, thus perpetuating the market distortions created by established, monopolistic interests (Trebilcock and Hrab 2005).

Public utilities in Canada thus remain part of a system of "managed competition" (Doern and Gattinger 2003, 5). A future area of contention is likely to be the extent to which the public utilities might continue their historical role as virtual extensions of provincial government public policy. Concerns about the environment now seem to act both as a vehicle for promoting the nuclear option and as a government rationale for supporting nuclear power. The CNA (2007) thus champions nuclear power as the means to both generate electricity and combat global warming. The National Energy Board (NEB) suggested in 2003 various scenarios to describe Canada's energy future, and in the "techno-vert" scenario nuclear reactors were regarded as a plausible option allowing Canadians to embrace both advanced technology and environmental conservation. Such claims of "clean nuclear" have not gone unchallenged. Some NGOs (e.g., Sierra Legal Defense 2006) have challenged the legality of such claims (whether they breach the Competition Act by making misleading claims). Others have disputed their veracity in general, arguing that they ignore known environmental impacts by selectively focusing on reactor output and not on the full nuclear fuel chain (Pembina Institute 2006).

Yet as noted above, contemporary critiques of nuclear power tend to focus on issues of economic competition and poor management (see Arai 2001, 416-21; Pembina Institute 2006, 75-78, 94-96). If nothing else, talk of restructured electrical markets in Canada has shifted to the forefront of arguments against the nuclear option the question of whether nuclear power can exist outside government support. Bratt's study of CANDU exports, for instance, concluded not only that every AECL foreign reactor sale has been accompanied by scandals about "payments" to the buyer (loans, bribes) but also that in general the federal government has heavily subsidized each sale (Pembina Institute 2006, 216, 227). Subsidies for domestic reactor projects have also come under fire, though estimates of both the extent of subsidies and their economic consequences vary widely.

An AECL-commissioned study by Ernst and Young (1993) claimed that 30,000 jobs had been created by the nuclear industry by 1992, whereas only $4.8 billion in subsidies to AECL had been granted up until then. Martin and Argue (1996) countered that this claim overestimated job creation by

40 percent and that the AECL subsidy (1952-95) was in fact close to $13 billion by 1995. Morrison (1998) also conducted a study of economic impacts, estimating $5.5 billion in the subsidy to AECL by 1997, but Martin (2003) claims that Morrison's study reproduces much of the Ernst and Young report. Martin thus claims that the subsidy to AECL (in dollars of the year) rose to over $17 billion in 2001 (1-6). Adams (2006, 1-2) recently placed the subsidy at almost $21 billion by 2005 and claimed that AECL is responsible for 12 percent of the current federal debt. The CNA (citing a report by the Canadian Energy Research Institute) recently claimed that nuclear energy constitutes a $5 billion per year industry, creating (approximately) 21,000 direct jobs and 10,000 indirect jobs and accounting for $1.2 billion in exports (2007, 1). This debate about economic worth promises to continue, with many critics claiming that nuclear power will not survive in a competitive market environment (see Adams 2000).

Canada has also experienced changes in its nuclear regulatory regime. In May 2000, when the Nuclear Safety and Control Act (NSCA) came into force, the AECB was replaced by the Canadian Nuclear Safety Commission (CNSC). For some, the CNSC is more efficient and independent than the AECB (Jackson and de la Mothe 2001). The CNSC concurs, noting that separation between a staff organization and a quasi-judicial tribunal creates independence (see the CNSC website at www.nuclearsafety.gc.ca). Certainly, staffing matters have improved over time, with staff numbers increasing from 50 in the early 1970s to over 500 by 2005 (Office of the Auditor General [OAG] 2000, Chapter 27, 13); the CNSC budget has also grown to $76 million (OAG 2005, Chapter 6, 3). Nevertheless, the OAG found that the CNSC continues the AECB non-prescriptive approach, relying "heavily on the knowledge and competence that its staff had gained earlier in their careers" in private industry (2000, Chapter 27, point 30).

This reliance on staff predominantly drawn from industry is the reason critics continue to label the CNSC "a 'lapdog, not the watchdog' of the nuclear industry" (Mehta 2005, 106). Technical expertise, embedded in a sympathetic bureaucracy, is thus said to continue a long-standing but largely invisible trend of bureaucratic agents making policy "behind-closed-doors" (Martin 2003, 7). Critics thus continue to argue that bureaucratic elites act to marginalize public involvement (Mehta 2005, 115), in part because the CNSC remains dominated by industry interests, but also because NRCan is dominated by pro-nuclear interests. Edwards (2005, 35) asserts that every minister of NRCan is "greeted by a phalanx of nuclear advisors ... drawn from the nuclear industry itself ... [so that] the government has in effect deprived itself of balance and objectivity." Indeed, the minister whom Edwards referred to, John Efford, had given a speech to the CNA in which he concluded that "nuclear energy must be a part of the Canadian energy process ... Everything we will do will be not only in our own personal interests, but will be in Canada's interests" (NRCan 2005). Bratt (2006, 73-74) similarly notes that over the past decade NRCan ministers have continually advocated nuclear power.

Even relatively detached analysis has concluded that bureaucratic support for nuclear power has ultimately "subverted public transparency and fostered government mismanagement" (Daniels and Trebilcock 1996, 6), leading to the recommendation that "nuclear energy policy deserves a full debate, not a closet discussion" (Doern, Dorman, and Morrison 2001a, 28). Of course, the track record thus far in regard to open discussions suggests that precautions must be taken lest power inequalities render participatory practices default exercises in public education (as shown in studies of open licensing hearings: see Mehta 2005, Chapters 5 and 6; Salter and Slaco 1981). In the 1996-97 public inquiry into a nuclear waste disposal concept (chaired by Blair Seaborn), the conditions under which participation operated limited the effectiveness of public involvement (see Durant, Chapter 5, this volume). Not only did AECL actively avoid addressing panel (and thus public) guidelines in preparing its disposal concept documentation (Durant 2007b), but also at the inquiry the public was marginalized via a narrow scope to the discussion, through a disposal proponent able to introduce substantial new material with little opportunity for public review and through the important discussions about implementing bodies, which were conducted outside the review context (Durant 2009b). Indeed, Lois Wilson, a member of the Seaborn panel, claimed that the federal government's response (NRCan 1998) ignored most of the recommendations of the Seaborn report (Canadian Environmental Assessment Agency [CEAA] 1998) and thus ignored public input (Wilson 2000).

Open discussion of nuclear projects has continued, in its historically truncated form, even as discussion of a nuclear renaissance heats up. The Ontario Power Authority (OPA), established in 2005 to forecast supply and demand and contract with the private sector for new generating capacity, has strengthened the nuclear option in Ontario. The OPA has recommended refurbishment of existing nuclear reactors and new builds, initially as an "excellent alternative" to other options (2005, 24-27) but later as a necessity lest the development of new electricity capacity be delayed (2007, 7-8, 20). Environmental groups claim that the OPA not only glosses over a host of problems with nuclear power but also continues to marginalize public input into energy policy decision making (see Sierra Club of Canada 2001, 18-23). If electricity market restructuring means "taking power away from monopolists and regulators where feasible and giving it to the competitive market" (Dewees 2001, 147), then it seems that the OPA's market guidance supports the "managed competition" thesis of Doern and Gattinger (2003, 5).

A Nuclear Renaissance?

Encouraged, no doubt, by the OPA position, OPG applied to the CNSC to build two new plants at its Darlington facility (OPG 2006), and Bruce Power applied to build two new plants at its Bruce County site (Bruce Power 2006). Both applications were approved to go to the licensing stage by the CNSC in summer 2008. This potential domestic renaissance follows an earlier rejuvenation of the reactor export market, with AECL having secured foreign sales to South Korea, Romania, and China since 1990 (Bratt 2006, 174-96). Claims that a new round of nuclear expansion will take place in Canada are now prominent features of public policy debate about the nuclear option. If the track record of policy making about nuclear projects is a reliable indicator, decisions about refurbishing reactors or building more will continue to exclude meaningful public input.

The issue of nuclear waste management assumes tremendous prominence in the context of nuclear renaissance discussions. Proposed solutions to the nuclear waste problem have the potential to become an obligatory point of passage for nuclear expansion plans. Indeed, various domestic organizations charged with solving the nuclear waste management problem (including Canada, Sweden, the United Kingdom, and the United States) have used their approach to nuclear waste management as a way to reinvent the face of the nuclear option per se (Durant 2007a). Canada's Nuclear Waste Management Organization (NWMO) is a prime exemplar of this trend. The NWMO has seemingly embraced its external political accountability (Durant 2006), recently proposing (NWMO 2005b) Adaptive Phased Management (APM) as a waste management approach. APM is meant to demonstrate industry social responsibility by apparently leaving key decisions about which options to take (e.g., storage or permanent disposal) and which implementation schedule to observe (how long to monitor, when to seal, etc.) to future political bargaining (Durant 2009b). The question that remains is whether there will be a public policy debate in which nuclear waste management is acknowledged as embedded in the broader policy of nuclear energy and in which such energy choices are treated as basic social choices.

To date, the NWMO has explicitly avoided engaging in this kind of debate, and this avoidance suggests that the democratic rhetoric the NWMO otherwise uses owes more to pragmatic necessity than to a full commitment to an open, democratic debate (Durant 2009a). Part of the problem with the NWMO is that the Nuclear Fuel Waste Act of 2002, which established the organization, drew on the "polluter pays" principle to populate the NWMO with waste owners and producers. This restriction of NWMO membership to waste owners and producers explicitly rejected the Seaborn panel's recommendation to form an arm's-length agency (CEAA 1998, 66-68). The NWMO can thus defer to its restrictive federal mandate to justify its neglect of the broader implications of nuclear waste management. The trouble with this stance is that the NWMO is comprised of powerful political-economic actors (e.g., AECL and OPG) with both a historically demonstrated (see Durant 2009a) and a currently active vested interest in turning proposals such as APM into support for nuclear expansion (see Durant and Stanley, this volume). These entities favour particular futures, and it would be naive politics to think that the NWMO somehow has absolutely no relation to the very entities that form it.

Alas, the more things change, the more they stay the same. This chapter has reviewed some of the contentious aspects of nuclear power in Canada, showing that concerns about economic merit, reactor safety, the health hazards of reactor operation and low-level radiation, and environmental degradation from uranium mining and other nuclear fuel-related activities have been constant features in public disquiet about the nuclear option. Throughout this history, public groups have complained about the lack of democratic scrutiny of nuclear projects and immunity from accountability of nuclear decisions and decision makers. The nuclear waste management issue shows every sign of continuing this lamentable trend. For instance, in June 2007, NRCan accepted the NWMO's APM, calling it a step toward ensuring a future for nuclear power (NRCan 2007), even though the NWMO had proclaimed that APM neither "promote[s] nor penalize[s] Canada's decisions regarding the future of nuclear power" (2005c, 20). The political duplicity at work here is actually the specific problem of limiting discussion about what technical proposals are for: that is, about the ends of inquiry. There ought to be full and open debate, in other words, about whether APM legitimates further nuclear expansion. Important social choices, such as energy options, deserve nothing less.

3

An Official Narrative: Telling the History of Canada's Nuclear Waste Management Policy Making

Darrin Durant and Anna Stanley

In this chapter, we argue that a number of organizations, loosely character-ized as the "nuclear establishment," have tremendously influenced the defining and framing of the nuclear waste management issue. We document the formation of a consistent and powerful characterization of the history of nuclear fuel waste management policy making. The pursuit of mutual policy directions by otherwise separate entities is not coincidental but fol-lows from shared goals with respect to the Canadian nuclear program and related concerns to normalize a number of policy directions. We suggest that the assumptions, goals, and policy aims of the nuclear establishment are continually being repackaged in and through what we call the "official" narrative. We show how this narrative obscures the partiality of this policy process, thus making it unwise to rely on that narrative as an innocent his-torical account. There is a pressing need for a sufficiently nuanced historical account of nuclear waste management in Canada, which we seek to provide here through a far less sanitized account than that provided by government and industry.

By the nuclear establishment, we mean an influential constellation of interests, including nuclear research and development corporations such as Atomic Energy of Canada Limited (AECL); regulators such as the Atomic Energy Control Board (AECB, now the Canadian Nuclear Safety Commission [CNSC]); nuclear energy utilities such as Ontario Hydro, now Ontario Power Generation (OPG); government ministries such as Natural Resources Canada (NRCan); and private industry corporations (including suppliers, manufac-turers, and mining companies). This constellation has been relatively suc-cessful in selecting the themes, perspectives, and questions that frame the nuclear fuel waste management issue. Although not a completely unified political constellation, it is a relatively stable set of pro-nuclear interests that has arrived at a shared position in relation to nuclear waste management in Canada (Durant 2009a; Johnson 2007, 83-86; Stanley 2006, 196; 2008, 65). Although we do not assume that these entities form a single actor, we argue that grounds exist for treating them as effectively functioning as a single actor in relation to nuclear waste management.

Furthermore, we argue that, because the Nuclear Waste Management Organization (NWMO) is comprised of and funded by waste owners and producers, efforts to separate the NWMO from this constellation of interests lack credibility. The ways in which this nuclear establishment tells the policy history of nuclear waste (through the official narrative) conceal their institutional agency. The official narrative takes shape in and through a number of publications and reports by industry and government bodies. They include AECL's (1994a) Environmental Impact Statement (EIS) with regard to a nuclear fuel waste disposal concept, NRCan's (1998) response on behalf of the federal government to the recommendations of the nuclear fuel waste management and disposal concept environmental assessment panel (known as the Seaborn panel), and NWMO consultation documents (2003, 2004e, 2005c). With reference to these documents, we will identify the events out of which the official narrative has been formed.

We identify five main features of the official narrative and offer our own critical reinterpretation of these narrative strategies. First, the official narrative implicitly endorses an appearance of smooth policy flow. Policy stages in nuclear waste management are presented as a relatively continuous and rational progression from an objective problem (i.e., that waste exists and must be managed in some fashion) to a socially acceptable and technically feasible solution (e.g., deep geological disposal, and then Adaptive Phased Management (APM), as explored through a rigorous technical and public review). We recast this smooth policy flow as akin to a process of *unfolding* in which initial technical and policy preferences have been repackaged as the exigencies of the situation demanded. Second, both AECL and the NWMO propagate an image of themselves as dutiful delegates of both federal will and international consensus. Each organization does this in slightly different ways and for different purposes: AECL to legitimate its proposed waste disposal concept of 1994 (Durant 2007b), and the NWMO to re-establish industry control over nuclear waste policy making (Durant 2007a; Stanley 2006). We complicate this dutiful delegate image by reinserting consideration of institutional interests, which include an interest in expanding nuclear power. Third, in the official narrative, internal policy machinations are pushed to the background, whereas an image of the nuclear industry fulfilling its social responsibility and obedience to government is pushed to the foreground. We invert this narrative strategy by bringing to the foreground salient instances of internal policy machinations. Fourth, the official narrative presents technical research and policy conclusions as pointing in the same direction (ultimately, permanent geological disposal). We do not deny that science and politics have been used to legitimate this direction, but we deny the spurious political and economic neutrality granted to this political direction. Fifth, industry actors take for granted that "waste owners" (prominently, AECL and OPG) ought to be the central performers in the management of nuclear fuel waste. We treat this as a normative preference deserving of more serious political consideration than has taken place; indeed, we suggest that this administrative structure undermines the credibility of policy making by tacitly privileging particular political and economic interests.

The Official Narrative

The AECL EIS (AECL 1994a) was the first statement on nuclear waste disposal intended for wide distribution. We thus begin our account of the official narrative with that EIS. AECL began the EIS by discussing the necessity of a solution to waste disposal before citing an AECB regulatory document (AECB 1987, known as R-104), which had declared that AECB's "preferred approach" was permanent disposal with no intention of retrieval (AECL 1994a, 2). Several government commissions were then noted to have "also concluded that disposal is necessary," (2) thus establishing a policy impetus for developing a waste management solution. The Hare report (Aiken, Harrison, and Hare 1977) was quoted as recommending "that waste should not be allowed to accumulate indefinitely in interim storage" (2). The Porter Commission (Porter 1978) was quoted to the effect that "there is clearly an urgent need to develop ultimate disposal facilities to ensure that these wastes are isolated from the world's ecosystems"(2).

This impression of policy consensus and momentum was reinforced when AECL quoted parliamentary committees (e.g., Brisco 1988), which "also recognized the need for disposal, stating that, whatever the future of nuclear energy, the waste it has produced must be disposed of" (AECL 1994a, 2). International consensus among "most nations" (3) also reinforced that permanent geological disposal was the optimal waste management option. Positioning itself at the forefront of this emerging consensus, AECL noted that its own 1972 committee of waste owners (AECL, Ontario Hydro, and Hydro-Québec) had concluded similarly (3). AECL noted policy support for this preference in the form of a 1974 joint federal-Ontario policy statement.

The NWMO treads a similar path to that of AECL, noting the long-term "recognition" by the "international community" that nuclear waste is best "buried and sealed deeply in stable geological environments" (2003, 16). Explicitly developing the dutiful delegate stance, the NWMO notes the 1972 joint waste owners' suggestion of deep geological disposal, the 1974 federal-Ontario policy approval, and the supporting recommendations of the 1977 Hare report and 1978 Porter Commission (17). Both AECL and the NWMO refer to these reports as the grounds for the establishment of a joint Canada-Ontario Nuclear Fuel Waste Management Program in 1978. Energy, Mines, and Resources (EMR, now NRCan) assigned AECL the responsibility for developing a deep geological disposal method, with assistance from Ontario Hydro and other organizations.

The NWMO (2003, 17) follows AECL in citing the 1987 AECB R-104 as having "confirmed deep geological disposal as the preferred approach in Canada," thus granting early regulatory legitimacy to the concept of deep geological disposal. AECL's EIS constructs a more detailed regulatory and policy background (1994a, 2, 307, 336). A 1981 federal-Ontario statement is cited to the effect that both site selection and implementation authority would not proceed or be decided on until after the disposal concept itself had been accepted (EMR and Ontario Energy Minister [OEM] 1981). AECB regulatory documents feature prominently. AECB (1985), known as R-71, is cited for its "position" that geological disposal was the preferable technical option. AECB (1987), known as R-104, is cited to the effect that techniques and designs that do not rely on human institutions were advantageous because these institutions were unreliable beyond several hundred years. The official narrative thus constructs a harmony between technical and political preferences.

Yet to avoid the harmony being viewed as the work of insider politics alone, the official narrative takes control over how the reader conceives of the boundary between the technical and the political. NRCan does this in a straightforward fashion, dividing policies that directed AECL to investigate deep geological disposal in 1978 from the 1989 decision of the EMR minister to refer a nuclear waste disposal concept to a full environmental assessment. The two stages are represented as recognition of "the importance of nuclear fuel waste management and the public's interest in this issue" (NRCan 1998, 3). In effect, a subtle differentiation has taken place within the official narrative. The harmony between technical and policy preferences at first reflected internal government-industry collaboration: wise policy preferences structuring scientific research. NRCan reconfigured the science-policy relation as of 1989, with autonomous scientific research now subject to external political review. Indeed, AECL's EIS had maximized the extent to which the EIS was prepared in response to environmental assessment panel guidelines and public consultation (AECL 1994a, i-iii), though critics disputed this claim (Durant 2007b). The NWMO has continued the emphasis on external review, presenting the 1996-97 public inquiry of "unprecedented scope, duration and cost" (2003, 18) as a natural political development in which the industry was a willing participant.

An important feature of the official narrative is thus the effort to minimize the perception of objections to external review, thus fostering an image of an apolitical waste management policy process. Yet in relation to the public inquiry, for instance, AECL consistently pushed for a narrow technical review (Murphy and Kuhn 2001, 258-60). Deleting such institutional agency from the official narrative serves to obscure links between nuclear waste disposal and preferences and goals relating to the future of nuclear power in Canada. AECL's EIS promulgated this image of minimal institutional agency by retreating behind the public inquiry mandate, which limited discussion to a concept alone and not future energy options (Canadian Environmental Assessment Agency [CEAA] 1998, 84-85). The EIS is thus silent on the consequences of accepting waste disposal (implementation notwithstanding). Indeed, AECL sought to isolate waste disposal from any consideration of political implications. Although acknowledging a "parallel inquiry" (into Canada's future energy mix) had been promised to the public by EMR when the terms of reference were released in 1989, AECL labelled as politically irresponsible those criticizing waste disposal because of their commitment to a nuclear phase out (Greber, French, and Hillier 1994, 64). The NWMO continues AECL's apparent innocence about future energy decisions, stating baldly that "our study process and evaluation of options was intended neither to promote nor penalize Canada's decisions regarding the future of nuclear power" (NWMO 2005c, 20).

Such marginalization of the policy implications of waste disposal allows for a narrow focus on questions about disposal technology. This marginalization strategy was utilized in response to the Seaborn panel report. The panel (named after its chair, Blair Seaborn), formed after the public inquiry mandate was set in 1989, had consulted the public extensively. Scoping sessions had been held in 1990-91, an initial review of the AECL EIS for compliance with panel guidelines had been held in 1994-95, and the public inquiry proper took place in 1996-97. The Seaborn panel reported its findings in 1998 (CEAA 1998). NRCan used some phrasing from two of the panel's four key conclusions to restate the panel's conclusions as one: "From a technical perspective, the safety of the AECL disposal concept had been, on balance, adequately demonstrated for a conceptual stage of development but that as it stands, the disposal concept has not been demonstrated to have broad public support" (1998, 4). NRCan proceeded as if this meant technical matters were on a sound enough footing, but what remained was social acceptance of the disposal concept. Yet this differentiation of so-called "social" from "technical" marginalized political dissent by prescribing public opposition as opposition to the disposal concept alone, despite the Seaborn panel's report that opposition was more nuanced. Social acceptance was lacking, in part, because the industry intentions in regard to waste disposal were perceived by the public to be tied up with expanding nuclear power (Durant, Chapter 5, this volume).

We suggest that the federal government built this NRCan interpretation into the Nuclear Fuel Waste Act of 2002 (creating the NWMO). The legislation invoked the "polluter pays" principle to justify establishing an implementing agency controlled and funded by nuclear fuel waste owners and producers. Rather than take seriously distrust in institutional interests, the

NWMO was handed a mandate to develop a "socially acceptable" solution via consultation with the public. The NWMO has interpreted its mandate as the task of consulting the public with a view to arriving at acceptance of a technology. The NWMO thus interprets its mandate narrowly; we suggest that interrogating institutional interests themselves is not incompatible with ascertaining what the public might think of technologies. We thus do not deny that extensive public consultation exercises have taken place, but we assert that the object of those consultations reflects continuing marginalization of political dissent and stubborn refusal to accept that much public disquiet is about institutional interests rather than technology per se. Moreover, the NWMO mandate presumes that waste owners should perform assessment work, even though the Seaborn panel recommended that an arm's-length agency should be formed to make such assessments (CEAA 1998, 66). As we discuss below, the NWMO's numerous references to the need to secure trust are laudable, but they ultimately reinforce the presumption that waste owners can and should be the central players in nuclear waste management. This presumption is especially problematic when viewed in light of the fact that, in accepting the NWMO recommendation of Adaptive Phased Management (APM) (NWMO 2005c), the federal government treated APM as part of securing a future for nuclear power (NRCan 2007). This enactment of policy implications compatible with waste owners' economic interests indicates the superficiality involved in focusing on technological options at the expense of readily discernible institutional interests.

Nuclear Power and Nuclear Waste
Similar to the AECL EIS, the NWMO maintains that waste exists and must be managed regardless of the future of the nuclear industry and that waste management will have no bearing on this future. Such narrowness owes much to the NWMO mandate, which omits consideration of the fuel cycle and waste production in the management of nuclear fuel waste. Yet consistent with privileging considerations of technology over institutional interests, the NWMO selectively included in its final report (2005c, 393-97) future waste considerations to support the adequacy of technical and engineering proposals. Three of the four scenarios considered involved nuclear expansion, while the remaining nuclear phase-out option was considered a pessimistic scenario based on loss of public trust (393). Given that the NWMO thus used nuclear expansion scenarios to support technical options, there appears to be little justification for excluding from public engagement consideration of the impacts of expanded nuclear production. The question that should naturally arise here is that, if the NWMO can implicitly consider energy options in order to shed light on technical options, then why can it not similarly expand its mandate in order to shed light on questions of institutional interests and the energy options themselves?

Given that proposed solutions to the nuclear waste management problem have in fact been used to endorse nuclear expansion, the time has come when the political duplicity of claiming that one has no relation to the other ought to be set aside. Although the NWMO recommends a separate discussion on nuclear energy, such promises of a parallel inquiry into energy issues have been made and broken in the past (Durant, Chapter 5, this volume; Kuhn 1997). While political discussion of energy options has thus not come to fruition, technical work on nuclear waste disposal has remained a puzzle to be solved within the project of developing nuclear power. We know, for instance, that AECL has conducted work on nuclear waste disposal since the 1950s (Auditor General of Canada 1995, Chapter 3). Early 1970s work by AECL considered monitored, retrievable storage as a credible option, in part because of an interest in retaining fuel flexibility but also because permanent disposal was considered unproven. In Canada, the shift to permanent disposal, as most preferred, followed the Indian bomb test of 1974, enabled by Canadian assistance to the Indian nuclear program, and thus domestic concern over the diversion of spent fuel to military uses (Durant 2009a). By 1977, the basic multi-barrier concept of placing waste 500 to 1,000 metres underground, in some kind of canister and in secure plutonic rock formations of the Canadian Shield (a crystalline rock formation that stretches across most of Ontario, Quebec, Manitoba, and the Northwest Territories), had thus been outlined (Morgan 1977).

Considerations about nuclear waste disposal in Canada were never separate from planning related to nuclear power development (the first commercial reactors were the Pickering A reactors, which began operating in the period 1971-73). Yet the contemporary official narrative constructs the waste problem as an isolated hazard, with a history of policy positions recommending that the hazard be dealt with via permanent disposal. The context in which these recommendations were made is conveniently deleted, as if the implications associated with solving or not solving the waste problem are irrelevant to past discussion. For AECL and the NWMO, the past is thus reconstructed in the image of the prescribed present. For instance, both AECL and the NWMO invoke the Hare report to the effect that permanent disposal in geological formations, probably in Ontario because that province is home to most reactors, ought to be the preferred option (Aiken, Harrison, and Hare 1977, 46). Similarly, the Porter report's (1978) reiteration of this conclusion is also invoked.

Yet the official narrative glosses over long-held doubts about the independence of the Hare report, which was sponsored by EMR, and two of its authors were former AECL vice presidents (J. Harrison and A. Aiken). Moreover, the context for the Hare report was growing public concern over nuclear safety and a perceived urgent need to settle on future nuclear energy policy. The introduction to the report thus states that "the urgency of the national energy question, the importance of nuclear powered electricity generation as a contribution to Canada's future energy supply and the increasing public concern regarding the overall safety of nuclear power have made it essential for the government of Canada to formulate policies for the long term management of the radioactive products of nuclear powered generating stations" (Aiken, Harrison, and Hare 1977, 1).

The Porter Commission, established in 1975 to review Ontario's long-range plans for generating electricity (which AECL does note at considerable distance – measured in pages – from its policy summary [1994a, 57]), viewed nuclear waste disposal in a similar context of whether or not to proceed with nuclear expansion.

The policy response to these reports again suggests how closely linked were waste disposal and the future of nuclear power. A 1978 joint statement by the minister of EMR and the Ontario minister of energy, formally accepting geological disposal in the Canadian Shield, directed AECL to investigate waste immobilization and conduct geological field testing. AECL was to select a site by 1983, and a disposal facility was to be operational by 2000 (EMR and OEM 1978). The official narrative constructs a more sober history by claiming that the waste disposal program has been a "no urgency" affair (AECL 1994a, 345; 1994b, 45). Moreover, the early 1978-81 siting program encountered much local opposition (Durant 2009a; Murphy, this volume). Communities throughout Ontario (including Massey and Spanish), and especially in the North Shore area and surrounding First Nations, made it clear that they did not trust AECL and Ontario Hydro and thus regarded claims that exploratory drilling was limited to rock characterization rather than site characterization as lacking credibility. This distrust extended to provincial legislatures, with Manitoba passing legislation outlawing the permanent disposal of nuclear waste within its borders in exchange for hosting AECL's underground research laboratory (at Pinnawa).

EMR and the Ontario minister of energy responded with a second joint policy statement, announcing that no disposal site would be chosen until after the disposal *concept* had been approved (EMR and OEM 1981). The concept thus became relatively placeless, which even the chair of the Hare report of 1977, F. Kenneth Hare, would later regard as a "self-defeating" separation imposed by ministerial decision (1997, 56-57). Notwithstanding local opposition to perceived site characterization work, developments in Ontario politics also figured in the separation of concept from site. A minority Progressive Conservative government during 1975-81 created an avenue for environmental critique to reach parliamentary debate and led to cautious policy making (Durant 2009a). Our main point here, though, is that the official narrative is structured by the claim that public engagement has increased over time. Although this is true in the sense of formal and invitation-based consultation exercises, it is not true in the sense that the public has been presented with a concept isolated from geographical, political, and economic dimensions. Limiting the object of discussion to a concept lacking concreteness has sanitized public engagement. During the public inquiry, this lack of concreteness resulted in complaints about the difficulty of evaluating the environmental, social (Durant, Chapter 5, this volume; Wilson 2000), and ethical (Timmerman, this volume) *contexts* of the waste disposal concept. The Scientific Review Group (SRG; see CEAA 1995b) made similar complaints, and the NRCan (1998) response to the Seaborn panel report (CEAA 1998) acknowledged that lack of concreteness had presented difficulties.

Nevertheless, this relatively context-less approach continued into the NWMO final recommendations. Despite a legislative requirement in the Nuclear Fuel Waste Act, requiring the NWMO to evaluate each option for nuclear waste disposal in a designated economic region (to avoid problems associated with a hypothetical concept), the NWMO refrained from doing so in its final study (NWMO 2005c, 145). Instead, it included lists of a large number of "potential economic regions" for the implementation of each of the options and assessed each option within what it called "illustrative economic regions" (2005c, 145-49; also see Murphy and Kuhn, this volume). Although the NWMO thus justified ignoring public calls to discuss future energy policy by invoking fidelity to its mandate, this mandate was interpreted more flexibly when it came to the discussion of economic regions. Given the historical connection between waste disposal and nuclear power, inflexibility with regard to discussing energy futures but flexibility elsewhere suggests the protection of institutional interests.

Negotiating the Terms of Reference

Although we have reservations about what the object of discussion has been, we also disagree with the implicit claim that establishing the grounds of public debate has been an apolitical process.

Central policy actors have not always endorsed public engagement. The AECB has been consistently criticized for a pro-industry regulatory style, which is certainly compounded by having to rely on former private nuclear industry employees to fill important regulatory positions (Durant, Chapter 2, this volume). AECB regulatory documents in the 1980s (R-71 and R-104) were produced via internal AECB-industry consultation and largely addressed engineering feasibility (Durant 2007a). In fact, the federal government noted that early 1980s AECB practice was not to consult widely and indicated that the AECB had been required to consult "those directly affected" since 1994 (NRCan 1998, 10).

Parliamentary bodies also followed the early AECB example. In 1988, the year before the terms of reference for the public inquiry were set, the Canadian Parliamentary Standing Committee on Energy, Mines, and Resources endorsed deep geological disposal and proposed that the concept be reviewed internally by the AECB, the Ontario Ministry of the Environment, and Environment Canada. The point to keep in mind in regard to such examples is that the NWMO has presented its public engagement efforts as a natural progression within the policy domain, yet these examples suggest that public engagement occurred despite the wishes of nuclear industry supporters rather than because of them. Indeed, Hare testified at the public hearings that one of the authors of the Hare report, Aiken, had left AECL "in disagreement with the management" over the "outrageous ... failure of the promoters of nuclear power to tackle this issue already." According to Hare, the nuclear industry had "neglected" waste disposal, did not take the issue "nearly seriously enough," and in fact "didn't see it coming." The impetus to take the issue seriously came from public pressure and insider dissent (1997, 80-82).

The official narrative also misconstrues the extent of political unanimity, for instance invoking the parliamentary report by Brisco (1988) to emphasize agreement on the "urgent need" assessment of the Hare and Porter reports (AECL 1994a, 2). Yet Brisco (1988) also stated that the disposal concept should be immediately reviewed *independently* of the nuclear industry and with public participation. Indeed, the report recommended a moratorium on nuclear construction until an acceptable solution to the waste disposal problem was found (37). AECL drew on its bureaucratic ally, EMR, to combat such sentiments. EMR had already bowed to AECL demands by pressuring the federal government to increase nuclear research and development funds, producing a report that extolled the virtues of the nuclear industry (EMR 1988a). EMR (1988b) thus responded to Brisco (1988) that solving nuclear waste disposal was integral to ensuring the future of nuclear power and that moratoriums on energy options pending public acceptance of clean-up practices would likely shut down most energy options.

Such nascent political conflict was also prominent during the setting of the public inquiry terms of reference, though the official narrative conceals it. Indeed, Murphy and Kuhn found that "there is no easily accessible record regarding the establishment of the terms of reference" (2001, 250). The final terms of reference instructed the Seaborn panel to "review the safety and acceptability of AECL's concept" but excluded from consideration the future of nuclear energy (CEAA 1998, Appendix A). The explanation for this narrow focus must acknowledge that it was not entirely due to successful efforts by pro-nuclear interests to isolate waste disposal from more contentious energy policy issues. Both federal and provincial governance style, in the late 1980s, favoured non-interventionist energy policy (Durant 2009a). Federally, Mulroney's Progressive Conservatives had adopted a non-interventionist governance style, reversing the previous Liberal era of federal colonialism of the provinces. In Ontario, the Liberal government found that it had to balance frictions within its own party about how to pursue the twin goals of environmental conservation and economic growth.

Nevertheless, AECL still pushed for a narrowly based technical review of the disposal concept. As a public manoeuvre, this amounted to an "attempt to evade ideological conflicts over the goals of energy and nuclear politics" (Kuhn 1997, 40). Scoping sessions held in 1990-91 revealed that two-thirds of the written submissions to the Seaborn panel viewed the narrow mandate as a subversion of political dissent (Kuhn 1997). As Murphy and Kuhn (2001) showed, this limited scope resulted from the need to balance conflicting agendas among government agencies and among those agencies and AECL. AECL wished to concentrate on technologies and methodologies and to sideline social, economic, and environmental considerations. EMR wished to circumscribe discussion of energy policy, believing that it would complicate planning for nuclear power itself. These preferences clashed with interagency demands (see Brisco 1988), the need to accommodate Ontario legislation in regard to participant funding and consideration of alternatives, and public calls for inclusion in decision making. The shift to a broad public review thus came against AECL's preference and reflected a compromise with regard to competing agendas. The Scientific Review Group (SRG), appointed to provide the Seaborn panel with technical advice, personified this compromise. Rather than function as a parallel (technical) review to the inquiry, as AECL wished, the SRG reported to the panel. Given this recent history of prominent nuclear-related organizations attempting to quell public involvement and discussion of energy policy, and given that some of these organizations now own and fund the NWMO (e.g., AECL and Ontario Hydro [now OPG]), we find it politically disingenuous for the NWMO to pretend that it is just a dutiful delegate of federal will when it claims to obey its mandate by avoiding discussion of energy policy. Indeed, the entities comprising the NWMO had a hand in setting that very mandate.

The Seaborn Panel and the 1996 *Policy Framework for Radioactive Waste*

We thus know that the mandate of the Seaborn panel contained elements, such as the requirement to address a broad audience rather than just scientific and engineering points of view, that conflicted with the institutional preferences of the nuclear industry. Moreover, while the neglect of energy policy reflected broader political currents, it remains true that pro-nuclear interests found such neglect in their strategic interests. A downturn in the economic fortunes of the nuclear option by the late 1980s meant that the hyperbolic estimates of industry growth, prominent in the 1970s, were no longer politically feasible in public. Put simply, the nuclear industry had been forced to accept that it could not present nuclear waste as a puzzle to solve within the project of advancing nuclear power, as it had done in the 1970s, because nuclear power itself was on the defensive much more than it had been in the 1970s.

If the dutiful delegate stance gains credibility because pro-nuclear positions became less abrasive in the 1980s (compared with the 1970s) (Durant 2009a), close research nonetheless reveals a combative nuclear industry. For instance, AECL appears not to have followed the guidelines of the Seaborn panel in preparing its EIS of 1994. The Canadian Environmental Assessment Panel (CEAP) had released guidelines in 1992 for the submission of AECL's EIS, but initial public comments (received August-November 1995) on whether the EIS of 1994 was in conformity with those guidelines indicated that most groups thought it was not (CEAA 1995a). Both public groups and the SRG concurred that the credibility of AECL's EIS was compromised by a lack of independent peer review, and many public groups argued that the EIS was written in response to AECB regulatory guidelines (R-71 and R-104) rather than panel guidelines (Durant 2007b). Indeed, an AECL advisory panel had complained to the Seaborn panel at this early stage that it was "disturbed" by the "indecisiveness" of "public policy" and frustrated because it was already "persuaded" by the research conducted (AECL 1995, 3-4). Technical potential thus constituted a sufficient ground to accept waste disposal. Such attitudes bring necessary balance to the NWMO's otherwise sanitized view of public policy, in which NWMO practices are said to be "fortified by the inherent wisdom of citizens" (NWMO 2003, 3) and in which the NWMO's APM proposal "advances a collaborative process in which citizens always play a legitimate role in making decisions" (NWMO 2005c, 4). Much of this talk appears pragmatic in intent when we note that the AECL advisory panel, in 2002, welcomed the establishment of the NWMO but declared its study process as yet another "delay" in siting that threatened to erode the knowledge base relevant to repository work (21-26).

The pragmatism of democratic rhetoric is also suggested by the way that the official narrative trumpets industry involvement in the public inquiry (NWMO 2003, 16-18; 2004d, 49; 2005c, 16, 21, 30) but underplays the role of internal industry-government negotiations. The most salient example of circumventing public involvement is the *Policy Framework for Radioactive Waste,* released in July 1996 but produced via industry-government consultation in the March-November 1995 period. The official narrative is so hesitant in regard to how these internal negotiations are presented that it contradicts itself. The policy framework is briefly noted in the NWMO's first discussion document, where the NWMO states that it "was meant to lay the ground rules and define the role of government and waste producers for the approach to waste management that was *anticipated* in the Seaborn panel report" (2003, 36; emphasis added). Yet the NWMO final report asserts that the federal government "*articulated* a policy framework" in "*response* to the report of the [Seaborn] panel," which led to the formulation of the Nuclear Fuel Waste Act (NFWA) of 2002 (NWMO 2005c, 154; emphasis added). The NWMO then notes that the NFWA enacts key components of the policy framework, including provisions for government oversight and the role of waste owners in funding and running the NWMO (330). We suggest that it is characteristic of the official narrative, rather than an accident of wording, that internal negotiations were first acknowledged as anticipating conclusions to be reached in a broader democratic forum but later presented as if they had been articulated in response to deliberations within that forum. That is, a central feature of the official narrative is the claim that nuclear waste policy making follows from and is directly responsive to public input.

The policy framework in question began its life as a 1995 Auditor General of Canada suggestion (Auditor General of Canada 1995, Chap. 3) to consult "major stakeholders" with a view to providing policy direction to the nuclear waste issue, given the time and money thus far invested. Documents that we obtained through the Freedom of Information Act show that consultation was limited to federal-provincial government agencies, owners and producers of nuclear waste (including AECL, Ontario Hydro, Hydro-Québec, New Brunswick Power, and other minor producers and owners of nuclear waste), private waste owners and industry actors (including the Canadian Mining and Energy Company [CAMECO] and Canatom, makers of nuclear technology), and lobby groups (e.g., the Canadian Nuclear Association [CNA]). In a briefing to the minister of NRCan, it was stressed that public exposure and consultation during the policy development process, especially with non-governmental organizations (NGOs), was to be explicitly avoided (McCloskey 1995). Discussion focused on a draft policy document, formulated by the Uranium and Radioactive Waste Division of NRCan, which asked for specific industry positions on the roles that the federal government, provincial governments, and owners of nuclear fuel waste should take in designing and implementing nuclear waste management plans (NRCan 1995). Unlike the restriction put on the Seaborn panel, this document frames these questions by relating nuclear waste policy securely to matters of future energy policy: "Resolving radioactive waste issues and progress toward disposal will make the nuclear option more acceptable as a source of energy, and reassure customers of the CANDU (Canadian Deuterium Uranium) reactor that Canada has a valid and integrated approach to the management of wastes from the CANDU fuel cycle" (1).

The end product of this consultation was the *Policy Framework for Radioactive Waste,* released halfway through the proceedings of the public inquiry itself. The framework stated that "waste owners and producers are responsible in accordance with the principle of polluter pays, for the funding, organization, management ... of their wastes" (NRCan 1996, 1). This policy framework did not so much "anticipate" the Seaborn panel as pre-empt and contradict it, which would recommend that an agency be established "at arm's length from AECL and the utilities" (CEAA 1998, 66). Although the "polluter pays" principle is commonly taken to imply that the polluter is responsible for the financial costs and associated technical work of pollution cleanup and prevention, the policy framework uses the principle to imply that the polluter ought to be the central actor in deciding how to interpret the framing of the issue, how to bring cleanup measures forward for consideration, and how to consult those that might be affected by or concerned about the issue. Moreover, as noted in 2000 by Lois Wilson (former Seaborn panellist), the policy framework compromised the Seaborn panel by suggesting that the public inquiry was window-dressing to behind-the-scenes negotiations. This allegation gains force when we consider that the government enacted the key feature of the policy framework, with only producers and owners of nuclear waste identified as potential members of a future agency (NRCan 1998, 7). Wilson thus argued before Parliament that the NFWA, in establishing (in 2002) waste owners and producers as comprising the NWMO, missed a "golden opportunity" to make a "fresh start" with an independent agency (Canada 2002b). Hence, the nuclear industry itself assumed a dominant role in setting the framework within which the NWMO would operate.

The official narrative further sanitizes the political field by deleting from the list of key Seaborn panel conclusions the recommendation to form an independent agency (see NWMO 2005c, 16). The official narrative thus operates with a rather constipated conception of independence. For instance, the NWMO final report constructs its democratic credentials via reference to "independent" entities, which include consulting firms, government oversight and regulation, experts, and its advisory board. Although we have some reservations about the extent of independence of some of these chains of association, in a broader sense their independence is largely immaterial. What remains unspoken in the official narrative, indeed obscured by the NWMO's reluctance to admit its parent organizations' roles in the setting of its mandate, is that talk of independence is really about how much discretion ought to be granted the nuclear industry in *any* decision making about waste management. In effect, the official narrative deflects attention to satellite bodies, when the real bone of democratic contention is the independence and specific role of the central body itself.

The Challenge of Aboriginal Narratives

The role of central bodies in decision making is clear when we consider the Seaborn panel recommendation that a process of participation with Aboriginal peoples be established and that Aboriginal peoples should design and execute the process (CEAA 1998, 66). The panel argued that AECL's EIS amounted to superficial treatment of Aboriginal peoples. The EIS noted that Aboriginal people (singular) are likely to be those most affected by the concept due to the overlap between traditional territory and the Canadian Shield. However, no attention was given to Aboriginal peoples' status or rights to that land (typically described in the EIS as "crown" land) or to moral and legal questions surrounding decision-making authority about it (AECL 1994a, 74). Instead, it was stated that, because Aboriginal people (singular) may be significantly affected "socio-economically," they will need to be specially consulted during siting and implementation of the concept (74).

During the public hearings of 1996-97, the testimony of Aboriginal peoples made explicit the ways in which they have been implicated in the landscape of the nuclear fuel chain. Compared with the narratives of proponents, those of Aboriginal peoples narrated radically different regional and historical experiences of the effects of that chain (Stanley 2006, this volume). According to the panel, the disposal concept itself, along with the consultation process about that concept, gave Aboriginal peoples cause for significant concern about their lands and rights (CEAA 1998, 20-21). AECL was judged not to have attempted to understand Aboriginal peoples' viewpoints (60). The disposal concept itself was judged to conflict deeply with many Aboriginal peoples' beliefs about their relationship with and responsibility to "Mother Earth" (CEAA 1998, Chapter 2). The panel also took itself to task for not having consulted with Aboriginal peoples in a fashion respectful of their cultures, languages, and consultative processes (CEAA 1998, Chapter 2). As we note below, the NFWA and subsequent work of the NWMO have not significantly addressed the panel's concerns regarding consultancy and respect (Stanley, this volume). With respect to the central role and waste owner composition of the NWMO, we note that the Seaborn panel flagged that such a role would be deeply problematic with respect to Aboriginal peoples.

The Concept of Social Safety

Neglecting the importance of an arm's-length agency also brings with it conceptual implications. Having spent the 1990-95 period in numerous scoping hearings and open house discussions, and having both acquired familiarity with other national programs for waste disposal and heard initial comments on AECL's EIS, the Seaborn panel heard testimony from over 500 groups and individuals (and read their written statements) in the 1996-97 public hearings. The Seaborn panel also witnessed the responses to public scrutiny of AECL, Ontario Hydro, NRCan, the AECB, the SRG, and various pro-nuclear industry groups. Drawing on this wealth of sources, the panel concluded that, "from a technical perspective, safety of the AECL concept has been on balance adequately demonstrated for a conceptual stage of development, but from a social perspective it has not" (CEAA 1998, 41).

The panel, fortunately, acted in accord with an abiding feature of democracy: be responsive to input from diverse constituencies. The panel thus interpreted its mandate in a fashion responsive to what it had been hearing during the inquiry. The panel interpreted the meanings of "safety" and "acceptability" in the same holistic fashion that many public participants used the terms, in which both are evaluated from the perspective of "social safety"

(CEAA 1998, Chapter 4; Wilson 2000). This interpretation implied that no simple line could be drawn between safety requirements and confidence in particular proposals. Nevertheless, NRCan responded that a management agency should develop a "framework ... relevant to an assessment of the options" and that "public acceptability of an approach ... will be arrived at by assessing waste management options using both technical and societal criteria" (1998, 11). NRCan thus set up a number of dualisms: framework/ assessment, public acceptability following an assessment, technical/social. This ingrained dualism deleted the import of the panel's recommendations (see CEAA 1998, Chapters 4 and 5).

The panel quoted the definition of safety in the AECL EIS (1994a, 63) as "meeting criteria, guidelines, and standards for protecting the health of humans and non-human biota." The panel then noted that both safety and acceptability were not absolute notions but relative to community standards and values. Safety, though, could be given a narrower definition than acceptability. Safety could include meeting regulatory requirements, being based on extensive participation in scenario analysis, using realistic data and models, incorporating accepted practices, being flexible, showing feasibility, and being subject to peer review. Yet safety is only part of acceptability. Acceptability was deemed to imply having broad public support, being safe both technically and socially, being the product of a social and ethical framework, being supported by Aboriginal peoples, being selected after comparison with other risks, benefits, and options, and being advanced by a stable and trustworthy proponent and checked by a trustworthy regulator.

The outlook of the panel was contained in the view that acceptability "designates" what constitutes safety. What it is to be safe cannot be known independently of what it is to be acceptable. Public acceptance and technical safety are intertwined. The federal government response implicitly divided the panel's interlocked concept of social safety into two, prescribing the task as achieving public acceptance of safety standards arrived at mostly via expert forums. In speaking against the passage of the later NFWA, Lois Wilson argued that the government was acting "as though the concept of social safety is invalid or unknown, and that all that is required is to convince the uninformed public of the technical safety of the proposal. This duplicity does not build confidence with the informed public" (Canada 2002b). For the panel, social safety and an independent agency were inextricably linked, for the acceptability of safety standards hinges on broad participation in and trustworthiness of an implementing agency.

Yet the official narrative suggests that the panel found the concept "safe from a technical perspective" (NWMO 2005a, 32). Indeed, where the panel referred to "social and technical shortcomings," the official narrative typically substitutes "social and technical issues" (see NRCan 1998, 14). In testimony to the Senate regarding the NFWA, Wilson made clear what the panel believed: "You will notice that our conclusion was filled with caveats, such as 'on balance' and 'a conceptual stage of development,' indicating that we were not at all convinced that the concept was technically safe. Indeed, 95 deficiencies in the technical proposal were documented" (Canada 2002b). We suggest that the official narrative conceptually separates the panel's technical objections from its recommendation to view those objections from the perspective of social safety.

Thus, in the NWMO background paper discussing the incorporation of the lessons of the Seaborn panel, the panel's concept of social safety is addressed separately from its noted technical shortcomings (2005d, 17-31, 44-110). The study forming the basis of the section on technical shortcomings, a report by OPG (2003) responding to comments about the disposal concept raised in the public inquiry, does not explore such comments as a means of applying a social safety perspective. In a subsequent study, an assessment team explored three management options (extended at-reactor-site storage, centralized storage, and deep geological repository) by measuring their scores against eight objectives (NWMO 2004a). *None* of those objectives was social safety. Hence, *in practice,* the panel's recommendation to join social and technical considerations in the concept of social safety has not been adopted.

In fact, the NWMO subtly undercuts the concept of social safety in Chapter 8 ("Safety from a Social Perspective") of its final study (2005c). The NWMO notes that it learned from the Seaborn panel the "need to give weight to both technical considerations and social and ethical considerations in the determination of a management approach" (154-55). It then outlines how its approach has been to develop an assessment framework that "integrates the broad range of social and ethical concerns with technical considerations" and factors them into "the objectives used in the assessment" (155). The NWMO admits that "social and technical notions of the core concepts of the assessment, such as what constitutes 'safety' and 'risk,' are so intertwined that they cannot be usefully separated for the purpose of the development and application of an assessment framework" (155). Nevertheless, it largely conflates social "safety" with ethical and social "considerations," which are presented as being brought to bear on technical considerations (implying less intertwining than the above statements would suggest). More importantly, the NWMO's use of social safety rests on a very different basis than what the panel envisioned.

The Seaborn panel conceded that members of the public differed about definitions and preferences and that securing representative community members was conceptually problematic. Yet the panel still thought that greater public involvement was always warranted (CEAA 1998, Chapters 4 and 5). That is, variation within the public was not a ground to avoid resting decision making about social safety on public consultation. In contrast, the

NWMO has destabilized "the public," rendering public preferences too diverse to form a quantitatively consistent basis of opinion conducive to policy making on the basis of categorized preferences (Durant 2006). In successive discussion documents, the NWMO has interpreted the reports of its independent consultants as revealing a fractured and difficult-to-satisfy public. Thus, even if the public may agree on broad societal goals, "disagreements are exposed when we chart a path to implement these goals" (2003, 4). Interpreting "social acceptability" to mean "public confidence" (2004e, 32), the NWMO argues that the common ground revealed by consultant reports (see Watling et al. 2004) parallels "a diversity of perspectives too" (2004e, 21). From our perspective, the NWMO constructs and implicitly prescribes the public as plural, internally differentiated, and riddled with unresolved conflicts.

By contrast, the NWMO constructs and prescribes experts as having stable and consistent scientific and technical judgments about safety and risk: "There is no technical evidence that would seriously question the technical feasibility or the long-term safety of a deep repository" (2005a, 35). Similarly, the NWMO makes the public and public knowledge plural, but expertise and expert knowledge remain unanimous and unitary (because they point in one unique direction, "the best"): "A socially acceptable management approach is one which has emerged from a process of collaborative development with citizens. It must take into account the best available knowledge and expertise, and be responsive to the values and objectives which are most important to citizens" (2005c, 17). This contrast between divided public and unitary expert means that social safety is used to suggest that a core base of technical knowledge can only be more or less compatible with social values. Whereas the Seaborn panel suggested that social safety be used to ensure that social values structure what is considered to be "safe" or "best practice," the NWMO inverts the concept in the way in which it uses social safety. That is, what experts categorically know to be safe is to be brought before public trial and made as responsive to diverse values and objectives as possible given the fractured nature of public preferences. We do not deny that this is a conception of social safety. Nevertheless, because the NWMO uses the concept of social safety in a different way from that outlined by the Seaborn panel report, we maintain it remains a different concept ("the meaning is in the use"). Pretending that it is the same concept grants the NWMO unearned democratic credentials.

The Nuclear Fuel Waste Act

Another facet of the Seaborn panel's concept of social safety was the injunction not to divide discussions of the future of nuclear power from discussions of waste management (CEAA 1998, 80-82). Only with the panel's mandate of 1989 were these two issues ever separated in Canada (Durant 2009a). The government response to the panel's report quickly reconnected them: "Taking steps to resolve the nuclear fuel waste issue would further support nuclear energy, and particularly the CANDU option, as a sustainable supply option for electricity" (NRCan 1998, 2). Negotiations from this point became internal government-industry collaboration. Draft federal legislation for the management of nuclear fuel waste and the establishment of the new waste management agency was circulated among OPG, AECL, and CNSC before cabinet saw the bill and before it went to Parliament for reading. Legislation was introduced to the House of Commons by a majority Liberal government in 2000, and Hansard transcripts for the two years during which this bill (C-27) was debated reveal that it provoked much debate from all opposition parties.

Prominent criticisms focused on the lack of a projected arm's-length agency, the lack of detail with respect to public and Aboriginal participation, poor provisions for public oversight, and the claim that public consultation would be about building social acceptance rather than incorporating diverse views. There was disappointment that nuclear waste was still being managed in isolation from public discussion on future energy policy.[1] Despite seventy-five proposed amendments to the bill by all houses of opposition, the NFWA was passed on 26 February 2002 (Canada 2002a) and given royal assent on 13 June 2002 (Canada 2002c). The NFWA establishes the relevant oversight mechanisms and reporting relations governing the NWMO and the NWMO's obligation to maintain trust funds to finance waste disposal. The NFWA directed the NWMO to investigate three waste management options (permanent disposal, storage at reactor sites, and centralized storage either above or below ground) and established that waste owners and producers would fund and operate the NWMO. An independent advisory board was also established within the NWMO. Much discontent with the NFWA centred on the fact that, as NRCan stated in parliamentary debate, the 1996 policy framework provided the "backbone of the legislation" (Canada 2001b). Indeed, representatives of NRCan described the 1996 policy framework as the "cornerstone" of nuclear waste policy making in Canada (Brown and Létourneau 2001, 113). In testimony before the House of Commons, those representatives described the NFWA as building on the policy framework and the Seaborn panel's findings and thus representing a "balance between the interests of Canadians and fairness to waste owners" (Canada 2001a). Nevertheless, NRCan representatives stated the four main parts of the bill were "fundamentally" based on the policy framework (Canada 2001a). Given that the NWMO's industry parents were instrumental in formulating the very policy framework that formed the cornerstone of the NFWA, we remain unconvinced by NWMO claims that avoiding energy policy is strictly a matter of government restriction. Currently, the NFWA lacks balance, we would argue, for it grants too much authority and responsibility to define and control nuclear waste management to waste owners with vested interests in expanding the nuclear option.

The Nuclear Waste Management Organization

The NWMO, established by the NFWA of 2002, seeks to manage the boundary between science and politics by making itself accountable and responsive to external authorities rather than isolating itself from them (Durant 2006; 2007a). The NWMO mission has been to "develop collaboratively with Canadians a management approach" (NWMO 2003, 6).

The NWMO claims that it is a "facilitator of dialogue" (2005d, 9, 39). This role obliquely acknowledges the Seaborn panel's advice that an independent agency is required because the nuclear industry is perceived as "secretive" and "self-interested" (13). Hence, the NWMO offers its role as "facilitator of dialogue" in aid of creating a "trustworthy *study process*" (9). Although this acknowledgment of the Seaborn panel's "arm's-length" recommendation is buried within a background document rather than the final report, we also suggest two problems with attributing this role to the NWMO.

First, the study process has not necessarily lived up to its promise (Johnson 2007; Johnson, this volume). For instance, despite the comments presented by Aboriginal groups to the NWMO, complaints have arisen that no proper consultation has occurred. The NWMO's Aboriginal dialogues targeted national Aboriginal organizations. Under the Canadian Constitution Act, neither these organizations nor the NWMO has the legal or constitutional authority to consult with Aboriginal peoples. Many of these organizations have pointed out to the NWMO that consultation must take place between individual Aboriginal nations and the federal government, for they do not have the constitutional, legal, political, moral, or territorial authority to enter into consultations on behalf of their constituent nations. Furthermore, organizations such as the Assembly of First Nations (AFN) have claimed that, despite their dialogues with the NWMO and the recommendations they offered, the work of the NWMO appears remarkably uncomplicated by their concerns, experiences and priorities (AFN 2005d).

Second, by its own standards, the NWMO study process is not trustworthy. The NWMO defines a trustworthy study process as one that is "transparent and the agenda for which is set by society at large" (2005d, 9). Yet NRCan based its response to the Seaborn panel on what in the panel's report was or was not consistent with the 1996 policy framework. That framework was also the cornerstone of the NFWA. The framework within which the NWMO operates is thus more responsive to internal government-industry discussions than to the broadly based Seaborn report. Nor is the NWMO mandate innocent of the institutional interests of organizations such as AECL and OPG.

As such, the process undertaken by the NWMO is in fact premised on explicitly ignoring "society at large." Society at large participated directly in the public inquiry of 1996-97. Preferences for an agenda explored by an independent agency, and an agenda in which energy policy is discussed in tandem with nuclear waste, have been unequivocally expressed by society at large. The boundary between the NWMO and its founding organizations is simply not great enough, and the influence that those organizations had over setting the NWMO mandate itself simply too large, for there to be any credibility in the NWMO's claim that it is just a dutiful delegate of federal will when it ignores energy policy and the demand for an independent organization.

Conclusion

On 14 June 2007, the government of Canada accepted the NWMO's recommended Adaptive Phased Management (APM) approach, declaring it an "initiative vital to the future of nuclear energy in Canada" and part of "steps toward a safe, long-term plan for nuclear power in Canada" (NRCan 2007). Although APM does incorporate repository monitoring and retrieval options, and does push decisions about various management options and an implementation schedule both into the future and onto political constituencies (Durant 2009b), the overriding intention is still to construct a deep geological repository and close it permanently. This intention has roots so deep in Canada's nuclear waste management policy past that the NWMO conceit of presenting APM as a *de novo* proposal born of recent public consultation lacks credibility. APM remains deep geological disposal repackaged for a political context in which accountability matters. We suggest that the official narrative is not incidental to this outcome but has been a crucial ingredient in the pro-nuclear game.

Note

1 See the *House of Commons Debates* between the first reading of 25 April 2001, through the second readings between 15 May 2001 and September 2001, and the subsequent third readings from 29 November 2001 to 22 February 2002. Also see Wilson (2000).

4

The Long Haul: Ethics in the Canadian Nuclear Waste Debate

Peter Timmerman

Compared with the substantial ethical, philosophical, and political debates in other countries over high-level nuclear fuel waste disposal (see Flynn et al. 1995; and Shrader-Frechette 2003), debates in Canada have been sparse. This is partly because a Canadian site – or sites – has not yet been singled out. All actors involved in this policy area acknowledge that the siting process will be socially and politically contentious (see Murphy and Kuhn, this volume), in part because of a long backlog of repeatedly expressed concerns. Up until now, discussions have been generated mainly under the auspices of the various commissions and reports. They have involved mainly industry statements, non-governmental organization (NGO) submissions, and, recently, a flurry of consultation activities by the Nuclear Waste Management Organization (NWMO).

In this chapter, I first review some of the historical debate on ethics in the Canadian nuclear situation. I then examine what I consider to be the greatest stumbling block in the debate: the question of the boundaries for what is to come under ethical review. Finally, I engage in some foundational work of my own concerning the underdeveloped, though often invoked, themes of trust and long-term burdens.

The fundamental social and ethical concerns triggered by the issue of nuclear fuel waste disposal are prompted not just by familiar debates over the rights and wrongs of siting waste facilities of any kind but also because, for many people, this particular issue raises basic questions about the trajectory of modern society. For them, nuclear waste is a "glowing" symbol of, or evidence for, things wrong with technocratic initiatives that generate risks as well as benefits (Beck 1992; 1998). The very question of what constitutes the scope and boundaries of ethical debate over nuclear waste is therefore deeply contentious.

But scope and boundaries are the critical ethical questions for those who have concerns about the various commissions and enquiries carried out in the area of nuclear waste management and disposal policy. For those who have these concerns, the constraints of the debate that exclude consideration of the larger social issues compromise virtually all the ethical proclamations and consultations issued over the years by "the powers that be" in the nuclear energy industry. These proclamations and consultations, however fine sounding they may be, are considered by nuclear opponents as simply a sequence of more and more sophisticated ways of getting participants to agree to the agenda of those in control of waste management policy.

This fundamental question of scope and boundaries makes it difficult to present a summary, let alone evaluation, of the ethical debate in Canada. Indeed, the reading and interpretation of the debate depend very critically on where one stands on this question and which ethical rules should apply to one's stance. For example, if one assumes a pragmatic, problem-solving stance, then a whole range of important value considerations about process, choice, compensation, and responsibility immediately becomes central. Thus, one can begin a process of mapping appropriate waste management or disposal sites, potentially approachable local communities, evaluating different ethical approaches to compensation (e.g., who, how much, how long, etc.). On the other hand, if one believes that these considerations are important only once one has addressed ethical considerations about the larger social, environmental, and economic context, then another, quite different, array of ethical concerns is central.

The United Church of Canada, for instance, raises serious questions about pragmatic problem solving in nuclear waste management. As it states in a submission to the Seaborn panel,

> we have arrived in this crisis because, in the late years of the Second Great War and the early years of a peace which included the Cold War and local wars that followed, the victorious industrialized powers decided they would use the same energy of the atom which had been used in the bomb to generate power, now "for peace." The decision had the immense appeal that a mixture of altruism and profit always has: but to it was added the immeasurable, mostly unconscious, appeal of pursuing the ultimate secrets of the universe, and, in a Godlike reversal of the entailments of "fall" and violence and war, using them for good and peace. So it was done.
>
> What was left undone at the time was any critical assessment of consequences, such as nuclear proliferation for military purposes, health dangers connected with mining and nuclear power generation, the hazardous drive to a plutonium economy, risks of terrorism, and, finally, the unsolved and perhaps unsolvable matter of the waste. Either because no thought was given, or insufficient thought, or because the technological hubris that has become endemic to the West gave its usual easy assurance, it was not so

much "decided" as simply and uncritically assumed that solutions to any later difficulties would be found in a timely fashion as needed. So it went ahead. (1996, 1-1)

This position tends to call into question the entire framework of modern technology and indicts modernity as a manifestation of the fall of humanity. Moreover, the dimensions of this particular critique condemn the very method that the pragmatic framework upholds. It is perhaps no wonder that the "pragmatists" and those concerned with broader social and ethical issues cannot "agree to disagree." There is nothing that they hold in common.

In this light, there is no single way to think (or talk) about ethics in modern Western societies. These societies may have certain forms of Durkheimean social glue that hold them together, certain tacit manners and practices that keep conflicts to a minimum, and certain legal histories and forms of appeal for general justice. But there is no agreed-on ethical touchstone (religious or otherwise) that can be appealed to in decision making. Into this vacuum, Foucault (1975) and MacIntyre (1984) have argued that assertions of emotional commitment and practices of power take command and are, in the absence of a common standard, the main factors in decision making (or decision stalling).

The general drift of ethical philosophy and practice in such societies has been, as a result, toward the conditions for equitable processes of public dialogue or deliberation. Equitable processes are seen as serving to address the difficult and irresolvable. There is a "weak consensus" that operates as a "floor" beneath which recent discussions do not generally fall and to which people generally subscribe if they wish to be taken seriously. This weak consensus centres on norms of truthfulness and reasonable politeness toward other participants in discussions and dialogues. From an ethical perspective, the weak consensus is that continuing dialogue is the good to which few object.

The Ethical Debate in Canada: 1991-2007
Some general ethical and moral positions are implicit in the earliest material on nuclear waste issues, such as the Hare report (Aiken, Harrison, and Hare 1977), Porter Commission (Porter 1978), and Interfaith Panel for Public Awareness of Nuclear Issues convened in 1984. Prior to the Seaborn panel, there appears to have been three significantly relevant discussions of the ethics of the waste chain. In March 1991, Atomic Energy of Canada Limited (AECL), in conjunction with Hardy, Stevenson, and Associates, conducted a workshop on "The Moral and Ethical Issues Related to the Nuclear Fuel Waste Cycle." Based on this workshop, Hardy, Stevenson, and Associates wrote two documents (1991, 1993). These materials had a clear influence on AECL's Environmental Impact Statement (EIS) of a concept for deep

geological disposal prepared for the Seaborn panel. The Institute for Research on Environment and Economy (IREE) prepared a third document (1995) that reports on a workshop devoted to the ethics of nuclear power and the waste issue.

The AECL Workshop (1991)

The ethics of nuclear power and the lengthier nuclear fuel chain were specifically excluded from the 1991 workshop. Furthermore, although workshop participants were fairly diverse – including philosophers, theologians, and First Nations representatives – most were from the nuclear industry or government.

Participants were asked to reply to a series of questions. These questions included the following.

Obligations to Future Generations
1 Should the rights of future and present individuals be weighed equally?
2 What are today's management obligations to minimize the burden on future generations?

Risk and Risk Sharing
3 What risks, related to the concept of deep geological disposal, are worth taking, given the probability of harm?
4 What can positively balance risk? Compensation?
5 Should some people (who receive no direct benefits) bear the burden of others, who do?
6 What is the responsibility of regions, provinces, et cetera, that have power stations in their own jurisdictions? Should they accept waste from elsewhere?

Consent
7 How should collective consent be obtained?

Uncertainty
8 What is the capacity of the social and institutional framework for managing nuclear wastes over thousands of years?
9 What is our uncertainty regarding the technology, and do we have enough information to make decisions?

Retrievability
10 Retrievability or irretrievability: Which is morally preferable?

The workshop's final report made a number of key points and recommendations. I summarize the report's conclusions (Hardy, Stevenson, and Associates 1991).

- The dominant perspective of the proponent was utilitarian – that is, seeking to maximize the good of society – though there were some "rights"-based elements (e.g., references were made to concern for the "most affected individuals") that might override such an overall good.
- Credibility and trust were seen as pivotal and a central problem for AECL. It was notable that the Berger Commission on the Mackenzie Valley Pipeline was cited as an example of a consultation process that broke out from the professional group of insiders.
- The workshop laid out various kinds of uncertainty (technical, social) that play different roles in the ethical/technical debates.
- It further suggested that "bought consent" is not "voluntary consent," especially when such consent involves poor communities.
- It also asked how we should weigh one proposed alternative that requires minimal management by future generations against another proposed alternative that requires more diligent management but gives future generations more flexibility.
- The workshop finally was agreed that "we have met our ethical obligations if we make the best possible decision today" (Hardy, Stevenson, and Associates 1991).

The issue that seems to have most bedevilled the workshop was uncertainty over the relationship between technical and societal risks and uncertainty over the means of reconciling the "rights of present and future generations." The industry, not surprisingly, although committing itself to a weak consensus on ethical norms and approaches, has taken as its main ethical stance the notion that "we have met our ethical obligations if we make the best possible decision today." A variation of the notion is that this generation created the problem and should do its best to come up with a solution. This is a pragmatic, solutions-oriented ethic and fits in well with what AECL saw as its responsibilities and its naturally preferred way of moving forward, given its priorities and its expertise.

The IREE Workshop (1992)
The other significant discussion was a session on nuclear risk during an IREE workshop that brought together many leading ethicists working in the area of nuclear waste management, including Lois Wilson, Gordon Edwards, Andrew Brook, and Conrad Brunk, as well as representatives from the nuclear industry. In particular, the papers by Brook (who would be a participant in the later NWMO Roundtable on Ethics) and Brunk (who would be a presenter to the Seaborn Commission) can be seen as early philosophical contributions to the debate. Brook's presentation (1995) sketched out competing philosophical positions and began to develop a sophisticated cost/benefit analysis that brought into the calculus enriched principles of equal worth and fairness,

together with freedom from costs, freedom from harms, and restrictions on the freedom of future generations to pursue their lives as they would live them. Brook's conclusion was that "we have an ethical obligation to find a permanent, passive solution to the problem of radioactive waste management. A true ethical skeptic or someone with a strong interest in ignoring the problem of nuclear wastes might still try to wiggle out, but to refuse to accept such arguments as ethically binding would be pretty much to get out of the business of justifying courses of action, finding good and sufficient reasons for what we do, altogether" (1997, 129).

Brunk (1995) took a "wider context" position. After reviewing the 1991 workshop, he said,

> the important point here is that a risk management decision and a risk communication strategy were adopted that imposed a decision on society *after the fact* ... The panel of ethics experts called together ... in effect were told by industry, "we have put this nasty waste on the doorstep of society because of decisions made long ago to pursue the nuclear option. Please tell us what would be the morally best thing to do?" Of course, the ethics panel could have been excused had they replied to the industry, "You have no right to put us in the position of having to extricate you from the moral impasse created by your own previous unethical decision." (1992, 2-3)

As has been noted, it is not clear that these two positions have anything to say to each other: the first accepts the given problem and attempts to work through ethical positions to a solution; the second considers the "givenness" of the problem to be the problem. This is particularly problematic when the industry refuses to promise to stop generating the problem.

I have christened this problem "Day 2 Ethics," where the original situation ("Day 1") is not subject to ethical scrutiny, but everything subsequent to it is subject to all kinds of ethical norms and value statements. This approach to ethics dovetails nicely with one familiar "progress-oriented" variant of the pragmatic position: What is the problem right here and now, and what is a good solution? Considerations of the past – according to this approach – are messy and guilt ridden and unlikely to produce a useful solution: better to stride on regardless.

The Seaborn Panel (1996-97)

The Seaborn review process produced a flurry of documents, some already cited, some by the proponents of the concept of deep disposal (e.g., Greber, French, and Hillier 1994) and a number by NGOs, ranging from First Nations communities to the Royal Society of Canada (these are unavailable at the present time). Outside the important proponent statements, which reiterated the pragmatic "solution in our time" stance, the two most extensive NGO statements covering extensive philosophical grounds were those of the United Church of Canada's Program Unit on Peace, Environment, and Rural Life (1996) and the Canadian Coalition for Ethics, Ecology, and Religion (CCEER) (1996). The CCEER report, based on a working group in consultation with the membership of CCEER (a network of citizens, theologians, and philosophers), and chaired by me, produced a manual of questions and extensive review essays drawing from a variety of ethical and spiritual traditions. For its part, the United Church submission focused on the fact that its membership was itself concerned in different ways about the nuclear waste issue (given that some of its members worked for the nuclear industry in Bruce and Pickering), and therefore it was thought that deeper theological questioning was in order. Importantly, one of the outcomes of their questioning seems to have been adopted by the NWMO in its final report, and that is the judgment that there should be no premature adoption of any solution. The United Church document cites at least six reasons for concern:

1 the inadequacies in the current Environmental Impact Statement;
2 the risk and uncertainties of the science related to the burial option;
3 the importance of retrievability of the wastes;
4 the burden imposed on future generations if serious problems develop with the burial;
5 the inappropriateness of severing concept analysis from siting and transportation; and
6 the premature and therefore irresponsible decision for geological burial at the present moment given the risks and the up to 100 years of storage time still available in the present management system. (1996, 4-1)

In addition to other recommendations, the United Church in 1996 requested that the panel ask the government of Canada to "hold the long-promised public review of the entire nuclear fuel cycle and its future" (1996, 4-2) – a consistent feature of NGO demands following from earlier government promises of such a parallel process. (In April 1998, in his appearance before the Standing Committee on Environment and Sustainable Development of the House of Commons, the chair of the panel, Blair Seaborn, stated that he had a number of times requested an extension of the mandate of his panel to do so, but this had never been agreed to [Wilson 2000]).

The final report of the Seaborn panel reinforced the weak consensus approach to future processes and, under what it called "safety and acceptability" criteria, stated that the concept must have things such as

1 broad public support;
2 safety from both a technical and a social perspective;
3 development within a sound ethical and social assessment framework;

4 the support of Aboriginal people;
5 selection after comparison with the risks, costs, and benefits of other options; and
6 advancement by a stable and trustworthy proponent and oversight by a trustworthy regulator. (Wilson 2000, 162)

The report gave no indication of what a "sound ethical and social assessment framework" might be. And, in its key conclusions, it made reference only to the notions that, although the technical safety of the concept appeared to be "adequately demonstrated," it had not been so demonstrated "from a social perspective," and that it did not have "the required level of acceptability" (Wilson 2000, 162). These phrases, which helped to trigger the next phase of the process, can charitably be described as "punting the ball down the field." It is clear that the panel was unable to cope with the ethical questions in spite of the many submissions and ethical concerns raised by opponents of the process.

The final report of the Seaborn panel also recommended that a nuclear fuel waste management organization be created at arm's length from the utilities and AECL so as to ensure a modicum of trust in the next stage of the process. This was a recommendation that was ignored as well in the government's response, which led to the Nuclear Fuel Waste Act (2002) and the setting up of the NWMO.

The NWMO and Ethics

The NWMO thus began in a compromised environment, with a long backlog of unaddressed ethical issues. There is not space here to review the extensive submissions and extraordinary range and variety of dialogue panels hosted by the NWMO over the length of the first part of its mandate (ending with its final report in 2005). Instead, I focus on the one explicit document on ethics produced by the NWMO process, entitled *A Roundtable on Ethics: Ethical and Social Framework,* the result of a series of meetings of a select group of Canadian ethicists (to which I was asked to provide an introductory report as part of the deliberations).

Before examining this document, however, it is worth noting that *Choosing a Way Forward: The Future Management of Canada's Used Nuclear Fuel – Final Study* (NWMO 2005c) contains the roundtable document as an appendix, underscoring its importance to the NWMO, but the NWMO does not dedicate a whole chapter to the topic of ethics itself (much of the ethics is implicated in various chapters, e.g., "Addressing Social, Economic, and Cultural Effects"). Section 1.2 of Chapter 1 does, however, deal with ethics specifically. It invokes the "sentiments and values of Canadian society": "This generation of citizens which has enjoyed the benefits of nuclear energy has an obligation to begin provision for managing that waste. That is consistent with the 'polluter pays' principle. Used fuel already exists ... We should not bequeath hazardous waste to future generations without also giving those generations the capability to manage the waste in a safe and secure way" (18).

It further invokes a principle of "humility" based on uncertainty, "avoiding approaches that are irreversible or overly dependent on strong institutions" (NWMO 2005c, 18). This invocation underpins a consistent theme throughout the report, one that marks a change in the management perspective overall, which can be seen as adopting some of the language of its adversaries: the NWMO should be engaged in adaptive management and should be prepared to change tactics (though perhaps not overall strategy) in the time to come. Flexibility, not determination, is the new watchword. The whole discussion is introduced and framed by the invocation of "the long view": "We are contemplating designing and licensing a system to last for periods longer than recorded history. That could lead to paralysis ... Furthermore, the technology used to store nuclear fuel waste today is safe, adequate, and affordable for some period of time and there appears to be no imminent safety or environmental crisis forcing a decision" (18).

Continuing with this line of discussion, this subsection concludes by invoking previous and continuing citizen engagement and dialogue. Tellingly, carved off from all this, and situated after the "values" discussion, section 1.3, "An Important Question of Context – The Future of Nuclear Power" (i.e., removing it from the context in spite of the section title), refers to the "impassioned arguments" about the future of nuclear power and proceeds first to neutralize them and then to abandon any responsibility for responding to them: "In this report, the NWMO has not examined nor is it making a judgment about the appropriate role of nuclear power generation in Canada. We suggest that those future decisions should be the subject of their own assessments and public process" (NWMO 2005c, 20). "Context" in this context is not therefore contextually connected to the ethical discussions in any serious way, nor does the NWMO even engage the critical arguments of whether the future of nuclear power is relevant to what it calls "the study process and evaluation of options" (20).

When we turn to the roundtable document itself, which is surprisingly short (four pages), in expectation of receiving a more detailed framework with which to tackle the ethical and social issues, there is little to go on. To begin with, the paper says that it has "constructed a framework of questions" that are "ethical principles incorporated in the framework." It is hard to see how questions in themselves are ethical principles. The paper itself says that the questions "aim to identify basic values, principles, and issues" (NWMO 2005c, 366). How does aiming to identify these "values, principles, and issues" constitute such principles?

The paper does not distinguish between values and principles. Are they the same thing? Conventionally, "values" language is generally applied to potentially, but not necessarily, articulatable goods – this is to enable them to be compared and often translated into "interests" that can then be traded off in negotiations. "Principles" language is generally used to consider more stable and formalized rules and codes.

To add to the confusion, at one point there is a reference to "key ethical principles" being in boldface, of which one such principle (Q9) is "costs, harms, risks, benefits." How are they (if they are plural) principles? They are a mixture of techniques and concerns, not principles. Another boldfaced "key ethical principle" is "liberty" – no definition of it is given. The reader is given no assistance in working through these contradictions. One procedural question asks whether the NWMO is committed to the "best ethics." But what is that "best ethics"?

It is not clear from the paper who is being addressed by the list of questions here. A reasonable reading would suggest that it is the NWMO itself that is being questioned (the so-called framework is "designed to guide its deliberations and its ultimate recommendations" [NWMO 2005c, 366]), but does this mean that it is the NWMO's "basic values, principles, and issues" that are being addressed, or is the framework to apply to the parameters of the nuclear waste issue as a whole?

Not boldfaced, but suddenly appearing near the beginning of the paper, is an explicit statement of an ethical principle (or at least method). According to the roundtable's paper, the "goal is to find and implement an ethically sound management approach. However, if no ethically sound management approach exists, adopting the ethically least-bad option available to deal with existing and committed spent fuel would be justified." "Adopting the ethically least-bad option" is a back-door way forward, especially given that we have no front door to enter wherein we can judge the ethically best option. Given the obvious concerns raised by this kind of approach – what has top priority if the ethically least bad option is to be implemented? what is the sequence of goods that must be sacrificed in order to get there? – the reader (and presumably the NWMO) is given no further assistance.

In spite of this, the paper goes on to argue that the creation of new nuclear fuel waste "must be judged by the standard of full ethical soundness" and not just "least bad." What that standard might be is not suggested, but even if it were the position is strange in this situation. It is an example of the dilemma of "Day 2" ethics. We are not allowed to bring the same ethical framework to bear on previous sins, only on future ones. We are not allowed to bring equivalent ethical analysis to bear on how the problem got here in the first place, as a way of preventing this from happening again. We are allowed to deal with subsequent symptoms, not original causes. This approach flies in the face of the fact that past sins have a way of not staying past. This approach might be acceptable to some people if solving the nuclear waste problem had no influence on present or future decisions about further creation of nuclear power and more waste. But this is clearly not true. In the government of Canada's response to the Seaborn panel, it explicitly says that "resolving the nuclear waste issue will further support nuclear energy, and particularly the Candu option, as a sustainable electricity supply option" (cited in Wilson 2000, 127; see also Government of Canada 1998). To refuse to bring the same ethical strictures to bear on the past is – in this particular situation – to foreclose on the ethics of the present and future.

The paper simply asks another question: "A question that urgently needs to be addressed is whether NWMO is dealing simply with existing materials and those that will be created in the lifespan of existing reactors or also with additional spent fuel? And this is no less than the question: What will the future of nuclear power in Canada be?"(NWMO 2005c, 20).

Yes. And?

As noted already, the NWMO has also refused to consider this question. An interim conclusion at this point would be that, after what has become a forty-year process, the legitimacy of the whole enterprise, from an ethical perspective, remains in doubt.

The Problem of Entrusting
Ironically, given its various equivocations and refusals, the NWMO has been quite innovative and creative in at least one area, which does involve a foundational ethical reframing of the whole enterprise. That is the dimension of time – what it refers to in its final report (NWMO 2005c) as "the long view" and what I have called "the long haul." Unfortunately, the NWMO in this report does not do much with this: the influence of the time dimension is limited to references to planning for the foreseeable future and aspects of the precautionary principle. Earlier on, in its scoping exercise, and again in an imaginative way, the NWMO commissioned a paper from Stewart Brand, originator of the concept of the "Long Now" (a clock and a concept for considering the next 10,000 years), entitled "Thinking about Time" (2003); it engaged in at least one future-envisaging scenario workshop that brought the future of nuclear power into the discussion (Mary Lou Harley, personal communication, 2003; see Global Business Network [GBN] 2003). But little of this finds its way into the final NWMO document.

I want, in the last two sections of this chapter, to focus on two aspects of the consideration of time in ethics. The first is the problem of entrusting. The second is what I call "the ethics of the long term."

Luhmann (1979) suggested that trust is one way in which we try to cope with uncertainty over time, and Parfit (1984) initiated a range of philosophical considerations around various related strategies such as promising, vowing, and other acts to defy or engage the passing of time. Almost everyone involved in the nuclear waste debate has identified trust as a central issue and concern (see Durant 2009b, Means 2007; Pijawka and Mushkatel 1992; and Rayner and Cantor 1987), including how difficult it is to generate and sustain trust.

If trust is critical, then the question of who or what we entrust our fates to is vital. Because the nuclear waste issue involves at least 100 years of active management and tens of thousands of years (possibly even a million or more) of containment, the traditional bonds of entrusting in our society are stretched to the breaking point. The whole purpose of the NWMO is to find a way of integrating scientific, social, political, and ethical demands such that it will both satisfy government and industry and earn the trust of the wider citizenry, some of whom may be future members of a "willing community."

The grave situation faced by the NWMO is that it entered the scene having to deal with a problem – nuclear waste – associated with an industry – the nuclear industry – whose origins, history, and occasional mishandling of aspects of power generation have made it appear singularly untrustworthy. Moreover, part of the legitimacy of government in a technocratic era rests in its professed ability to manage technology rationally. The rise of the modern bureaucratic state is crucially based on the premise that utilitarian calculations mixed with expert projection can anticipate and cope with situations as they arise. On the occasions when the symbols of managed rationality have broken down – Three Mile Island, Chernobyl, Vietnam, Iraq – they have threatened the legitimacy of the entire system of governance. Indeed, one of the elemental problems with the nuclear issue is that it is indelibly associated with Hiroshima and Nagasaki, and they constitute what Lifton (1982) refers to as a "broken connection": a possible foreshadowing of a complete break in the continuing narrative of human experience on Earth due to supposedly rational planning. "Nuclearism" raises in a foundational way the question of the long-term narrative of our life on Earth. Nuclear waste is inevitably part of that question.

As a result of this reckless, supposedly rational, proliferation of "risk" (Beck 1992), we find that the opposition (e.g., from environmentalists) to these symbols finds itself arrayed on the side of "prudential" and "precautionary" behaviour with regard to proposed future risks, while governments, administrative experts, and corporate elites are "risk prone" – in spite of their claims to prudence and precaution. Under these circumstances, the opposition to processes and organizations such as the NWMO has as one of its premises that a continuation of these kinds of processes and organizations is not just a symptom of the problem but also the problem itself. The system that created the nuclear waste created the NWMO.

The tragedy, therefore, is that, given this history and the underlying distrust of the nuclear industry, there is nothing that the NWMO can do to make itself trustworthy. Only a radical break with the system can originate and sustain trust over the long term. To the opposition, trust cannot be retroactively bestowed by good behaviour.

Ethics of the Long Term

I would argue that we require something more than what we've seen in the history of ethical debate about nuclear waste in Canada. I would suggest that part of that something more is to begin to consider how the very long term influences our ethical approaches, not in terms of process, but in how we envisage the long-term future, how our lives and the lives of others will play out in the future, and how that envisaging situates and creates different ethical stances.

The NWMO is committed to a belief in incrementalism: that is, a belief that we should keep moving forward the way we are for however long it takes for the waste problem to be solved. We should come up with a preferred solution and deal with problems as they arise thereafter (this is its "adaptive management" approach). This incrementalism not only supports a continuing set of pragmatic goals but also provides the foundation for utilitarian cost/benefit analyses, forecasting, and other "reasonable" processes. These processes project the current world slightly improved into the near future (since this is what makes their calculations work and reassures us that things do not have to change too drastically). Building on this incrementalism, society will be able to make manageable trade-offs, ensure dialogue, and work toward reducing the local uncertainties that might provide obstacles to some kind of site selection.

For many opponents of this approach – whom we might call "foundationalists" – incremental solutions are a betrayal of what ought to be at stake when we are confronted by the problem of nuclear waste. These opponents of incrementalism wish the ethical discussion to be directed toward the case of nuclear waste disposal as a symptom of the maladjusted ways in which modern society thinks about progress, time, and responsibility. The foundationalists treat the length of time over which we have to be concerned about nuclear waste not merely as a technical problem but also as an indictment of our overall thoughtlessness about the long-term consequences of scientific and technical progress. To foundationalists, the fundamental trajectory of the society as a whole comes into question, in part because the contours of the overall risk strategy adopted by our society seem to become clearer.

Similarly, the length of time that the waste will remain dangerous does not just make values important: it changes the nature of the valuing. It is as if the new vision of distant lengths of time sobers some people up or makes them look at the issue from a "God's-eye view." If I can put it this way, using a mathematical image, as the time over which the burden of responsibility for dealing with the waste lengthens, the approach to considerations of ethics becomes more idealized, more absolute, more foundational as it approaches eternity asymptotically. We are required to stand back from the engine of today and ask questions such as "what is the nature of a society that builds nuclear reactors with forty-year lifetimes and thereby generates waste with thousand- and million-year lifetimes?"

However, these kinds of questions do not get the foundationalists off some ethical hooks of their own. For example, the refusal to deal with the waste problem on principle – first we must consider the future of nuclear power, and then we can consider the future of nuclear waste – in practical terms means that those who currently bear the burden of risk will continue to do so, in the stalemate between positions. Is this fair?

An obstructionist strategy designed to provoke the raising of larger social questions also carries other risks. We can see this in the global warming debate with the refusal by many environmentalists to consider "adaptation" to global warming as an option on the ground that it will undermine struggles to "mitigate" global warming – that is, if we say we are going to adapt, then we have thereby given up the struggle to do something to stop it. However, should attempts at mitigation fail (as appears to be increasingly likely), this may increase the potential risks and the burden of rushed adaptation. We thus see being played out on a planetary scale some of the agonies of nuclear waste opponents: the feeling that, as soon as one has accepted any of the working assumptions of the powers that be, the game is lost, yet meanwhile the engine – pragmatic, incrementalist, progressive, et cetera – drives on.

Returning from climate change to the nuclear waste issue, there is also a curious tendency by some nuclear opponents to invoke some new and even more magical future technology that will transmute the waste or somehow solve the problem. So let's wait (a variation of the "there's no hurry" stance). In this way, even some opponents of a "technocentric" approach to nuclear waste can find themselves caught in the dream of modern technology: disliking today's technology, they put their hopes in future discoveries. They are subject to the very critique that they themselves use against the rise of the nuclear industry: the working assumption that someday a technological solution is magically forthcoming.

These kinds of expectations about the future link to basic orientations or optimisms/pessimisms about different aspects of the future and where one places one's emphasis concerning future uncertainty. They may profoundly influence where one stands on an issue such as nuclear waste.

So, for example, behind the dream that tomorrow will bring us a magic fix is an unspoken optimism, not just about future technology, but also about the continuation of benign management into the future – someone will be there to apply the new magic. But what if there is no one there in that future or no one whom we want to be there?

Proponents of deep disposal argue that we should not expect future above-ground society to be benign and continue to manage the waste reasonably. The closest that opponents come to this topic is in the context of terrorism. Although proponents of deep disposal argue that it will put the waste out of the hands of terrorists, opponents of nuclear power argue that "solving" the nuclear waste problem will lead the way to more nuclear power, which will in turn generate more waste, which will be more easily obtainable by terrorists.

So we have people who are technological optimists about the present and technological pessimists about the future; we also have people who are managerial optimists about the present and managerial pessimists about the future; and so on. Depending on what you are defending against, and in which basket you are preparing to dispose of your eggs, these stances will help to shape your priorities and ethical stances.

Are we then stuck for the unforeseeable future? The long haul? Philosophically, we are betrayed at the moment by a gap in our ethical theories between incrementalism and foundationalism. We do not have an ethics for the long haul (what to do between a little after now and a week before eternity). This is not just a need for a "middle range" of ethics dealing with the stretches of time from 100-10,000 years; it is also a need to rethink the ways in which we do stretch our minds out into that future: we have few ethical bridges into the long-term future.

The literature clusters, as I have indicated, mostly into the immediate and near term versus the equivalent of the absolute. Even the literature that explores our responsibilities for future generations seems to devote much of its energy to questions such as whether there are real people whom we can identify and therefore feel responsible for, how much discounting is allowable, and so on. We can also find the incrementalist/foundationalist debate playing itself out in discussions over the validity and usefulness of the "precautionary principle" – Manson (2002), for example, argues that only a "catastrophe principle" can play the role that those who originally championed the "precautionary principle" hoped it would play. It appears that only the presumption of catastrophe might now make us change our incrementalist ways.

The gap is not to be filled, I think, with such rickety structures.

It is striking and not, I think, accidental that this gap mirrors a similar gap in the most powerful source of ethical structures in the Western tradition, our religious heritage. The Christian element of this heritage focuses on the one hand on the morality of immediate conduct, and on the other it focuses on the intersection of time with eternity, first with the arrival of Christ and – to come – the apocalypse. In between now and then, the metaphors and symbols (many of them involving nature) have usually assumed, in the words of the Bible, that "seedtime and harvest shall not fail" (Genesis 8:22).

The firmament will remain firm. Except at the end of time when the apocalypse will resolve all things. But this does not cover our current situation, where we are making decisions that affect the sky, influence the seasons, and determine which species will survive to the year 2100.

The literature on promoting the concept of "fiduciary trust" (Brown 1994) is one possible bridge with its considerations of those elements of social life – parks, valued documents, et cetera – that are deeded to society "in perpetuity." This reference to touchstones of longevity gives a hint about one way forward, recasting some of the ethical debate around long-term issues.

When we "shift gears" concerning future time, in our thinking and feeling, we begin to deploy different metaphors, symbols, and myths as we reach ahead. In the shorter term, our thinking and feeling may include forecasting, economic discounting, and the durations of our own lives as exemplary; at 25-50 years, our metaphors turn toward children, families, houses, trees; beyond 50 years, they turn toward grandchildren, parks, older trees, towns; beyond 100 years, we invoke eras, generations, tribes, nations, cities; beyond that, we speak of peoples, races, geologies. These stretches of metaphor – which I would have to argue for more extensively than I can here – underpin and evoke different forms of ethics.

Moreover, these movements of the mind – or the society – forward increasingly require the integration of the past. The strong point of the critiques made by the United Church of Canada (and others) is that history is barred (or selectively invoked) in the pragmatic orientation toward the future. The narrowing of the waste argument depends in part on the narrowing of the history and the geography of what brought us to this point. One elementary principle of an appropriate time ethic might be that *it should go back at least as far as it intends to go forward.*

My conclusion to this chapter is based on the notion – or more explicitly a research hypothesis – that a time-centred approach to ethics is valuable in reconsidering what kinds of ethics are appropriate for considering the high-level nuclear fuel waste problem. I think that it would recast both the risks and the responses to the risks in a different light.

Without some such approach, we may miss the fact that the metaphors, symbols, and imaginative timescapes from which we derive our ethical languages and structures are currently impoverished. It is not enough to make reference as some authors do to the need to take into account non-Western worldviews or to invoke the "seven generations" mythology of certain First Nations peoples, as if such references will by themselves extend our ethical imaginations. The nuclear waste issue provides us with a timescape that challenges the assumptions behind our ethics, as do other environmental issues with long time horizons: these issues remind us that our ethics cannot be disembodied in the ways that have grown comfortable. For example, a recent report on collaboration between Inuit elders and scientists about the increasing impacts of climate change (Community of Inuvik et al. 2005) points out that one of the underplayed impacts is the erosion of traditional elder knowledge that depends on hundreds of years of reasonably consistent seasonal rhythms. This knowledge is critical to traditional social structures, cultural practices, and ethical norms and is not separate from them. In other words, our short-term risk strategy (i.e., to buffer ourselves against climate variability by using fossil fuels) is destroying their long-term risk strategy (i.e., careful and insightful responses to nature's ways). The changes in the physical environment alter the ethics.

We lack a truly embodied ethics for dealing with the long-term threats to our survival that are looming over, making holes in, and warming the skies above our future horizon, not to mention burdening future generations with the detritus of our thirst for energy. Brown's (1994) work on trust and stewardship gestures in this direction, as does the "ontological imperative" of responsibility – "act so that the effects of your action are compatible with the permanence of genuine human life" (Jonas 1984) – a categorical imperative that contains both idealism and a pragmatic commitment to real-time concerns.

It is clear that the lengths of time characteristic of the nuclear fuel waste problem and any solution have challenged business as usual – which is why the NWMO was deemed necessary – but the challenge remains unmet. Deep geological disposal is what Jungians would call "the shadow" – a vast project, like some kind of inverted pyramid, whose lengthening shadow through time is cast by the glittering ephemeral life on the surface of our society.

This problematic relationship between time and ethics in our culture was perhaps best illuminated by an incident involving the Yakima Indians in Washington state, who found themselves in a workshop about the long-term disposal of nuclear waste at the Hanford site. The problem was raised about how to cope with possible visitors to the site 10,000 or more years into the future – whether universal language, appropriate signage, forbidding landscape structures, et cetera might be the best ways to warn those future unknown visitors. The Yakima spokesperson answered, "Oh, that is no problem. Just tell us where the stuff is, and we will let them know. We aren't going anywhere."

5

Public Consultation as Performative Contradiction: Limiting Discussion in Canada's Nuclear Waste Management Debate
Darrin Durant

Framing Inadequacies

Jurgen Habermas' concept of a performative contradiction (1973, 1990) points to ways of interacting with others that are similar to making promises with your fingers crossed behind your back. One such practice involves limiting public dialogue about democratically sensitive issues through exclusion from discussion or imposition of interpretation. This chapter argues that just such a practice characterized Canada's public inquiry concerning a nuclear waste disposal concept, held 11 March 1996 to 27 March 1997. It has remained similarly applicable to the public consultation efforts of the Nuclear Waste Management Organization (NWMO) since 2002.

The inquiry's mandate limited discussion to a waste disposal concept and excluded from discussion considerations of the future of nuclear power and energy policy in general (Canadian Environmental Assessment Agency [CEAA] 1998, Appendix A). The mandate had been established in 1989, largely through government-industry negotiations (Murphy and Kuhn 2001). Even the panel established to run the inquiry, chaired by Blair Seaborn, found the narrow scope of the mandate regrettable (Wilson 2000, 11). The inquiry was divided into three phases, with phase one discussing social and ethical issues related to nuclear waste management and the criteria by which safety and acceptability ought to be decided. Phase two focused on technical issues, and phase three consisted of nation-wide community visits. In this chapter, I focus on the phase one debates about the scope of the public inquiry itself, drawing on testimony offered at the inquiry.

In making such testimony readily accessible,[1] this chapter aims to illuminate why public groups at the inquiry were dissatisfied and what frustrated their expectations. Moreover, utilizing actual testimony as evidence is consistent with a normative commitment to actual deliberation as the cornerstone of deliberative democracy (see Habermas 1990, 15, 66-67, 94, 211; 1996, 147, 151, 450). I show that public disaffection resulted from procedural restrictions on allowable discussion. These restrictions denied the relevance of both the institutional interests of waste disposal concept supporters and the future implications of decisions about waste management. Restricting the scope of political deliberation reinforced the pre-existing institutional influence of technical and policy elites and exacerbated already strained relations of trust. Indeed, phase one revealed contrasting social realities of trust and control, which extended beyond distrust and variable influence. Invoking the untrustworthiness of "the other" constructed and subsequently prescribed the character of that other. For the public, prescribed others were posited as most likely to constitute future decision-making communities. Historically based assessments of dominant actors thus informed judgments about what could be expected from future implementing bodies. Different assessments about future involvement in the policy process also influenced what different groups wished to see occur, based on their judgments of trust. Established policy actors were more confident of future involvement and could thus be content with marginalizing specific arguments of their critics. Non-governmental organizations (NGOs) and other public critics were less confident of their involvement in future decision making and thus attempted to marginalize both specific policy actors and their arguments.

Although the public inquiry is now in the past, an analysis of the problems that befell that inquiry promises to offer valuable lessons for the future of nuclear waste management policy making. Nuclear waste management remains divorced from energy policy, and the NWMO talks about the trustworthiness of managing institutions but does so without taking seriously the Seaborn panel's recommendation that an "arm's-length" agency was needed to handle waste management matters (CEAA 1998, 66). Exploring the basis of discontent at the public inquiry reveals that contemporary waste management policy making recapitulates past errors by failing to address public frames of meaning. These findings confirm Brian Wynne's general thesis about the causes of public disaffection with expert-led policy making. Wynne argued that two pernicious "framing" assumptions can bedevil deliberative practices (2005, 66-69). The first is an assumption that public groups are more concerned about downstream risks and impacts than upstream issues such as the human purposes driving research and policy. Phase one showed that critics distrusted the policy intentions and past practices of disposal proponents. Moreover, risk estimates were not separated from the groups providing them. The second assumption is that public meanings and issue definitions are often considered the domain of authoritative technical and policy institutions rather than lay citizens. Phase one demonstrated this assumption by the way in which a narrow mandate procedurally weakened what could be deemed the meaning of the concept.

I thus show that disagreements over the scope of the mandate facilitated divergent sets of discourses about the meaning of the deep geological disposal concept itself. Technical, industry, and bureaucratic groups constructed the concept as a flexible tool for indicating how a pressing problem could be solved *in principle*. Public groups rejected this meaning, regarding the concept as both strategically vague and carrying with it particular political and economic interests. I suggest that the restrictions on the scope of the mandate privileged the meaning of the concept already favoured by technical, industry, and bureaucratic groups. The public inquiry thus became a battle in which public critics contested the use of deliberation to implement an already decided outcome (the meaning of the concept) rather than to negotiate outcomes. This is not a claim to the effect that the inquiry was simply a "smokescreen." I am claiming instead that public groups wanted to negotiate the concept on offer and that the meaning of the concept is not synonymous with either its safety or its acceptability. It was a precondition to negotiating safety and acceptability, but the narrow scope of discussion both prevented full deliberation about what the concept meant and tacitly privileged an instrumentalist version of the concept.

In the concluding section of this chapter, I extend these points to a brief consideration of the NWMO's Adaptive Phased Management (APM) plan (2005c), which was accepted by the government in June 2007 (NRCan 2007). APM also rests on a strategic separation of concept from energy policy, and the NWMO continues to neglect framing issues even while it touts its own democratic credentials. Both factors plague hopes for the democratizing of science policy and prefigure continued public dissatisfaction with waste management policy making.

The Mandate
Prior to 1989, nuclear waste disposal represented a means to solve a problem within the ongoing project of commercial nuclear power (Durant 2009a). In both the Hare report (Aiken, Harrison, and Hare 1977, 1) and the Porter Commission report (Porter 1978, xiii, 95), waste disposal meant industry survival, even though the former recommended that nuclear expansion should not be delayed pending a solution and the latter recommended a moratorium on nuclear expansion without a solution. During the 1970s, Atomic Energy of Canada Limited (AECL) and Ontario Hydro directly connected the fate of nuclear power to solving the nuclear waste disposal problem, so much so that site assessment work in the period 1978-81 was halted (in part) because of this connection (Select Committee on Ontario Hydro Affairs 1980).

Although a joint federal-Ontario policy statement of 1981 declared that site assessment would not proceed until a concept for waste disposal had been approved, waste disposal as a concept remained connected to the fate of nuclear power. Throughout the 1980s, the federal department of Energy, Mines, and Resources (EMR) connected solutions to the waste disposal problem to ensuring that nuclear power had a future (1982, 1; 1988b, 15).

So did a prominent parliamentary report, even if it revived Porter (1978) in recommending a moratorium on nuclear power without a solution to waste disposal (Brisco 1988, 37). Nevertheless, when the inquiry mandate was announced in October 1989, a wedge was driven between a waste disposal concept and energy policy implications. Discussion of energy policy was prohibited. What was once irredeemably connected was now separate.

This political divorce owed much to a decade in which federal policies had shifted mid-decade, from the Trudeau Liberals and their interventionist policies in favour of subsidizing megaprojects to the Mulroney Progressive Conservatives and their non-interventionist policies in favour of deregulation and reduced subsidies of megaprojects (Durant 2009a). Ontario had changed too, having emerged from the Progressive Conservative dynasty of 1943-85, which had become rabidly pro-nuclear. The Liberals, toward the end of the decade, found themselves divided over how to pursue both economic growth and environmental conservation (Durant 2009a). Hence, both policy climates favoured a public inquiry that would not force their hand on large-scale (interventionist) energy policy, especially because the 1980s represented a period of stagnation for nuclear power (Durant, Chapter 2, this volume).

This broader policy context notwithstanding, it remains true that AECL pushed for a narrow technical review of the waste disposal concept. The shift to a broader public review represented a balance of interests: AECL and EMR preferred to avoid broader implications and issues (because they had led to opposition in the late 1970s), Ontario wished for discussion of alternatives, and it was widely perceived that a broad forum was required for such a contentious issue (Murphy and Kuhn 2001). Indeed, EMR implicitly acknowledged that energy policy implications were involved by promising a parallel review of energy policy (Murphy and Kuhn 2001, 262). This promise was not sufficient to prevent public groups, during the scoping phase of 1990-91, from criticizing the narrow scope of the inquiry (Kuhn 1997). This critique became more refined by the time AECL submitted its Environmental Impact Statement (EIS) on the waste disposal concept in 1994 (AECL 1994a). NGOs in particular, in initial reviews (submitted 1994-95) of whether the EIS was in conformity with the Seaborn panel's guidelines of 1992, argued that AECL treated the review as if the concept was only being subjected to an engineering feasibility study (Durant 2007b).

Not only was the mandate contentious before and during the public inquiry, but it continued to be evaluated differently on its completion. Some have argued that violations of the inquiry mandate, such as raising the energy policy issue, thereby abused the democratic process (Robertson 1998; and see footnote 7 below). This position appears to confuse democracy and authoritarianism. Critics are closer to the spirit of democracy, arguing that stifling debate about energy futures implicitly endorsed the continued production of irradiated nuclear fuel (see Edwards 2005, 20). In normative sympathy with the critics, I suggest that the mandate operated during the hearings as an implicit mechanism to control and discipline the very citizenry that the participatory mechanism of the inquiry was meant to include.

Public Meanings

Natural Resources Canada (NRCan) opened the inquiry by claiming that "it is important that we start to determine preferred positions on implementation issues and look at the implications."[2] When asked what happened to the promised parallel review and whether a nuclear phase out was being considered,[3] NRCan cited provincial objections to federal intrusion and reservations about the usefulness of another review as leading to a 1991 decision to defer a parallel review.[4]

Critics soon signalled that the concept was not considered separate from energy policy by asking whether the mandate was going to be expanded accidentally or intentionally.[5] Seaborn responded that the panel was "as unhappy as you are that the governments have not seen fit to meet this earlier commitment."[6] AECL and Ontario Hydro were certainly careful to avoid linking the concept to the future of nuclear power. Indeed, this denial of a connection simultaneously limited the opportunity to debate alternative energy futures. AECL's R-Public, which accompanied the EIS and addressed public involvement and social aspects (Greber, French, and Hillier 1994), had in fact used such surgery as a resource to combat opposition to both waste disposal and nuclear power. Although acknowledging the promised parallel inquiry, AECL claimed that opponents of nuclear power used opposition to waste disposal as a means to phase out nuclear power. Storage apparently kept radioactive waste a visible political burden (Greber, French, and Hillier 1994, 64). Nuclear insiders offered colloquial versions of the R-Public accusation: "Some of those opposing the disposal of nuclear wastes may be doing so as a tactic to oppose nuclear energy in general so that they can claim 'there is no solution to the waste problem.'"[7]

AECL maintained the conceptual surgery by dividing "opinions" (the section in the R-Public within which the quotation below appears) from more "factual" concerns: "The question of how widely nuclear power is utilized, or how provinces choose to generate electricity, is not something that can be answered by the waste disposal concept. The concept is designed only to provide a means for the safe disposal of the waste that has already been produced, continues to be produced, and may be produced in the future" (Greber, French, and Hillier, 64).

AECL's EIS maintained the decoupling by reducing waste disposal to an instrumental concern with technical problem solving: solving waste disposal was an urgent need regardless of the future of nuclear energy (1994a, 2). Ontario Hydro reiterated the instrumentalist defence, testifying that nuclear expansion was a separate issue and that it would "live with the outcome of the hearing"; indeed, its only concern was "to manage the waste that we have in a safe, environmentally, socially acceptable and financially responsible way."[8]

Nevertheless, supporters of the disposal concept did not universally sustain either the decoupling of waste disposal from nuclear power or the division between opinions and facts. Some collapsed the facts into opinions by assuming that opponents' assessments were biased by their opinions. Thus, it was said that the disposal concept possessed "good technical integrity," whereas those opposing it did so for "political reasons" that might ultimately compromise safety.[9] Others separated facts and opinions temporally, presupposing a time in which public opinion had given consent to nuclear development, after which only technical questions about implementation remained. Thus, the Atomic Energy Control Board (AECB) argued that society had already made its choice to use nuclear energy, and this meant that the present situation was best conceived of as factual inquiry into what to do rather than opinion-based debate about energy options.[10]

Others, such as the Canadian Nuclear Society, dispensed with subtle political decouplings of nuclear power from waste disposal altogether: "Acceptance of the concept will ensure that there is a safe and technologically sound method of used fuel disposal. This will, in turn, permit Canada to continue to develop and utilize nuclear power as a safe, environmentally friendly, reliable, and cost-effective means of generating electricity, for the benefit of all Canadians."[11]

Some industry groups linked their approval of the disposal concept to job creation within the industry and a greater "confidence that the nuclear option is sustainable and environmentally beneficial."[12] By means of the claim that "everyone benefits from nuclear energy," offered as an argument in favour of accepting the waste disposal concept,[13] any pretense that energy policy was not being implicitly *made* via waste disposal acceptance was left behind. Similarly, the Canadian Nuclear Association (CNA) responded "yes" when critics asked whether its support of the disposal concept related directly to its support of the continuation of the nuclear energy option.[14] NRCan also linked nuclear waste to nuclear power, claiming that it was prudent public policy to plan for concept implementation because the government was committed to nuclear power.[15]

Critics of the waste disposal concept were more uniform in always connecting the concept to public policy, often by focusing on the indecision apparent among supporters of the waste disposal concept. Concerns were thus expressed that potential commercial profit may lead to Canada becoming a repository for foreign nuclear waste. Despite NRCan's denial of plans to import foreign waste,[16] NGOs rather comically rebutted the denial, citing a TV interview in which an AECL vice president stated that waste retrieval clauses enhanced the prospects for export sales.[17] When the CNA defended the AECL representative, on the ground that sometimes a representative makes claims not universally held,[18] NGOs concluded that bad corporations might simply taint their members.[19] Other NGOs cited precedents, such as the government's reversal of a commitment not to accept "imports" to the hazardous waste facility at Swan Hills (Alberta) after siting,[20] as grounds not to trust decision makers.

Beyond interrogating hidden political preferences, critics claimed that institutional laziness and inertia characterized nuclear public policy making. The danger in allowing waste disposal was that it may be taken to imply tacit approval to refurbish commercial nuclear reactors or build more. Regardless of the political preferences of waste disposal proponents, critics thus insisted that the implications of waste disposal plans needed to be subjected to democratic scrutiny.

What's the Agenda?
Critics thus sought to define the boundaries around the deep geological disposal concept: Was the agenda waste management or nuclear industry expansion?[21] Possibly, the panel's mandate was significantly out of touch with actual political manoeuvring: "Clearly people who are knowledgeable about these issues and are invested, to whatever degree, in the nuclear industry are making longer term (and different) plans for a proposed high level radioactive waste dump than are considered in the mandate for this Panel. If this is the case, then this Panel has a responsibility to take on a larger mandate, to consider all aspects of these issues, and to report on this to the ministers who are responsible for this process."[22]

This mixing of questions about motives, purposes, and democratic accountability largely pre-existed the public inquiry. Critics predicted as early as 1986 that separating a concept from both a site and nuclear power was the most likely political strategy for the nuclear industry to adopt. A concept could thus be approved with little fuss and acceptance used to legitimate producing more waste (Poch 1986, 206-7). Indeed, the very idea of nuclear waste disposal *meant* nuclear expansion, for industry endorsed deep geological disposal "to keep the nuclear industry alive" (Robbins 1986, 170).

The public inquiry thus represented a means for critics to scrutinize the human purposes that they regarded as embedded within the deep geological disposal concept. For some, this was AECL's efforts to sell more reactors.[23] Critics were often frustrated that such purposes had to be coaxed out of supporters of disposal. One NGO reminded the panel that independent parliamentary inquiries had found that AECL assessments were "biased by its commitments to nuclear power."[24] Another showed critics' frustration by using the question-answer format of the inquiry to get the CNA's intentions (yes, it favoured nuclear expansion) on record.[25] Another argued that, to the extent that approving the concept was going to be taken as approval for nuclear expansion, each project had to be discussed.[26] It was this potential for one mandate to beget another that defined what the disposal concept meant for critics.

The pregnant potential of concept approval intersected with distrust of possible implementing organizations (AECL and Ontario Hydro), especially because power disparities between local communities meant that an implementing agency could easily marginalize opposition and co-opt a community.[27] Concept approval also threatened complete loss of influence over the decision-making process: "We have to do something with this stuff, and yet as soon as we do something with it we give the green light to make more."[28]

Citizen Science?

Does all this mean that political rights alone were at stake? For instance, few public critics grounded their claims solely by reference to research science. Many in fact invoked their ignorance of research science, but this ignorance does not mean that "science disappeared" in the inquiry exchanges. Indeed, during phase two, critics would prove extremely adept at identifying technical deficiencies in the concept and making arguments about the lack of satisfactorily defined technical parameters to the concept (Durant 2009b). What needs to be understood about public opposition is that critics disputed the meaning of the science on offer, indeed the very meaning of the issue in dispute at the public inquiry. Thus, invoking ignorance of science was part of efforts to manage the boundaries between ignorance and knowledge and between expertise and politics. Indeed, critics occupied a broad spectrum of claimed knowledge of research science, but all managed to articulate conceptions of the issue in dispute and what was at stake.

For some, science was irrelevant to an appreciation of hazards: "I don't have to be [a] scientist, you don't have to be a scientist to know what ... the effects of this nuclear waste is going to be on the environment or to all living life."[29] For others, worries remained despite ignorance: "Port Hope, mom. No knowledge, no brains, no science here, I'm just kind of worried about my kids."[30] This absence of technical knowledge nevertheless pointed to the importance of other roles in deciding issues (e.g., parental responsibility). Often the absence of technical understanding was appealed to in order to argue that technical groups were not performing *their* proper roles. Thus, some considered that alternatives had not been properly explored: "I am not a scientist. Not being one, I cannot judge the merits of these different transmutation technologies, but AECL was supposed to in your [Seaborn panel] guidelines, and it didn't."[31] Claims that technical groups were not "doing their jobs" were common: "I don't know what the answer would be [to waste disposal], but that is up to the people who are qualified and should be able to come up with a better answer than burying it."[32]

Science was either deemed redundant to common sense or implicitly invoked as a role not properly being performed. Others drew on their own experience to offer locale-specific knowledge. Several First Nations groups drew on their familiarity with Canadian lands to challenge assurances of low risk. Some did so via intuition-based claims: "They [Aboriginal elders] have this sense we cannot ignore."[33] Others cited firsthand experience of bays that became murky in the space of a generation: "In terms of science and scientific terminology, that is what they call experimentation over time or study over time."[34] Sometimes Aboriginal identity and experience were invoked to recommend that research science be conducted in a fashion that "incorporates our perspectives, our values, and our traditions."[35] Other times the intent was to directly challenge the usefulness of research science: "We know our land better than anybody else in this country. There is nothing that any scientists can tell us about what's on our land."[36]

Claims about technical matters that were continuous with acknowledged expertise in some fashion (e.g., experience-based claims that could contribute to an overall picture) were better received than claims offered as discontinuous entirely (e.g., claims citing political rights as the basis of superior knowledge). Technical consultants acknowledged that "people from in the North" can assist in risk assessment and uncertainty analysis because of "generations" of familiarity with the North and an ability to help scientists "brainstorm" "what we haven't thought about yet."[37] Some public critics were of a similar mind, shown when a First Nations representative was criticized for attempting to insert Aboriginal spiritual accounts into the environmental evaluation as "factual claims" rather than "a basis of values."[38] NGO representatives thus tended to emphasize their "local discrimination," constructing their contributions as continuous rather than discontinuous with orthodox technical evaluations. Northwatch spoke of its experience in northern communities, which had become a "nuclear sacrifice zone" because of economic vulnerability and mining companies that practised "flood and flee." Northwatch requested better, more trustworthy science, attuned to local impacts and demonstrating social responsibility. Local familiarity could contribute to that project, for it was tempered by "a certain cynicism or scepticism with respect to the nuclear industries that perhaps other regions don't have."[39]

Constructions of self that performed distinctions between the individual speaker and science in general were thus means to say that the salient issues were not "just scientific." A familiar trope was to preface comments with "I'm not a scientist, but I really think that perhaps I am speaking for a few people."[40] Even those who defended the necessity of some technical understanding admitted that other issues were at stake: "Environmentalism is some combination of a sophisticated value sense and an understanding of science."[41] Where the status of science was downplayed, this was rarely a matter of levelling scientific versus other considerations per se. Rather, the differences between scientific and other considerations were discursively reproduced in order to emphasize the relevance of a broad range of issues. Indeed, constructing differences between science and other factors prescribed the necessity of different roles in decision making.

Constructing Political Constituencies

Symmetrically speaking, competing political constituencies construct the other constituency and thereby implicitly prescribe the features of that constituency. For instance, former Seaborn panellist Lois Wilson (2000, 3) noted that the government had separated the deep geological disposal concept from siting because of fear of public opposition, even though the panel quickly realized that the public was not about to accept this division. Wilson's account suggests that there was a concern among governance groups to locate the right kind of citizen – a "reasonable citizen."

The AECB made clear its perception of the difficulty of locating reasonable citizens, noting that it had changed practices from decades ago, now embracing public consultation as a means to avoid being "locked into a particular paradigm." Yet the problem remained of "how to generate confidence [and] perhaps trust."[42] The AECB later pleaded with those attending the hearings to dissociate the AECB from the industry. The AECB divided groups into three: the first two either accuse the AECB of defending the industry or give out "deliberate misinformation" because they have a "goal," and the third is the "genuinely concerned citizen who really wants to find out. Those are the people that we have to talk to, find out what we are not communicating, find out what their concerns are and do our best to address them."[43] The AECB thus implied that its harshest critics were not the audience that it needed to convince, nor were such critics seemingly out to represent matters fairly.

Other technical groups made similar divisions between citizens whom they thought could be satisfied and those who could not, the latter being a "faction" that "mistrust[s] any technology or engineering."[44] The same group later doubted "the efficacy of involving lay persons in the development of waste management action plans" because of the variable reception of technical reports (lay citizens and media ignored them, whereas "opposition lobbies" pored over them and requested more).[45] This schema prescribed the public as either interested but ultimately factional luddites or not interested and not worth consulting.

Government groups also noted both the importance and the difficulty of public consultation. NRCan noted that the time has passed "when one simply can have technical experts walk in and impose a solution of virtually anything on a community."[46] The Nuclear Energy Agency of the Organization for Economic Co-operation and Development (OECD) proclaimed that deep geological disposal was a balanced opinion of public plus specialist but, when queried, admitted that the OECD had relied on consulting experts. The difficulty in consulting the public included the fact that the public largely rejected proposals coming only from experts.[47] The political constituency that formed the target of consultation efforts was thus routinely constructed as difficult to access, relatively fickle, and possibly belligerent. Some otherwise favourable toward the public still constructed it as emotive, describing a low-level waste repository project as "not a science project, this is not an engineering project; this is a project about dealing with people and dealing with human emotion."[48]

Others suggested that public education campaigns would mean that a better-informed public would support the proposal.[49] Both attributions assume that a deficit in understanding the technical issues was the cause of rejecting the disposal concept. Some industry insiders were far less diplomatic in their constructions of the public. Some bluntly divided those opposed to the disposal concept on the grounds of "false premises and bullshit" from those not explicitly and publicly outraged and thus actually having no "concern" at all.[50]

NGOs and lay citizens opposed to the disposal concept similarly sought the right kind of citizen: a "reasonable" or "true" scientist in this case. It was acknowledged that scientists could get it wrong, as in "DDT and Thalidomide," but mostly because of the perversions of "economic reality."[51] Given expected limits to public knowledge, trustworthy experts were needed. Social scientists called to testify advised that experts would have to be independent, without agendas, and sensitive to how the public understands risk to be trustworthy.[52] Many argued that "if science is independent of the operator it can be trusted."[53] Ideal models of independence as expertise freed from money and power were thus common. Sometimes philosophically general models were deployed, such as notions of a "true scientist" who practises "repeated verification of any conclusions."[54] Other times a more philosophically specific notion was used, such as the Popperian conception that science works when "every researcher is motivated to disprove the concept."[55] Experts per se were thus not the problem; their connections with untrustworthy institutions were mainly the issue. Thus, it was suggested that Whiteshell Underground Research Laboratory (Manitoba) personnel, who conducted much of the work relevant to the disposal concept but were "not associated with AECL," could form an implementing organization.[56]

Constructing the other also meant both prescribing how the other should behave and figuring out a basis on which assessments and judgments could be trusted. The federal government made clear that it wished to be viewed as a prudent planner, one that could act ethically in regard to future generations as well as responsibly in light of sustainable development ideals.[57] The AECB made clear that it saw itself as having to make judgments in the face of uncertainty. The only way to move forward under such conditions was to inject "a fair element of common sense" into proceedings.[58] AECL and Ontario Hydro argued that they had to be trusted because "the public expects us to move ahead towards disposal. Ontario Hydro supports geological disposal. With certain qualifications, several review groups have indicated that geological disposal is viable. Safety assessments have been carried out in other countries confirming the safety of geological disposal. Deep disposal is a policy direction in most OECD countries."[59]

AECL also outlined five principles that any implementing organization ought to adhere to: safety and environmental protection, voluntarism, shared decision making, openness, and fairness (1994a, 256). These principles could carry an organization into the future plus demonstrate sensitivity to public input.[60] Yet critics were not about to base their *assessments* of trust on the claim that moving forward *would* require trust and that AECL and Ontario Hydro *could* be trusted because they would adhere to particular principles. The past, not the future, was for critics the relevant domain for deciding whether or not to trust them. Here is where agenda talk mattered, for critics were not about to trust organizations with such a vested interest in a particular future. As one NGO argued, "you don't start science with your conclusion and then build the evidence to match the conclusion. That is almost a hallmark of bad science."[61]

The Issue of Public Trust

Some critics regarded the entire science of waste disposal as untrustworthy because it was funded by the nuclear industry.[62] Others were more discerning, accusing AECL of being self-interested and thus not trustworthy.[63] Some claimed that Ontario Hydro had not saved the money it claimed to have collected to fund disposal,[64] which Ontario Hydro admitted was true in the sense that money had been used to retire debt rather than create a segregated fund.[65] Others argued bluntly that AECL was the kind of organization capable of bribing communities to gain acceptance for a disposal site.[66]

The "body language" of AECL and Ontario Hydro was consistently cited as indicating that they were not to be trusted. Aboriginal groups accused AECL of duplicity and bribery, complaining that visits to First Nations communities were vehicles for pitching claims about economic prosperity masquerading as information sessions.[67] Others claimed that Ontario Hydro "don't listen to anybody," especially "Indians [and] women."[68] Some participants spent their entire presentations discussing their negative experiences acquiring information from AECL.[69] One book about such difficulties carried the not-so-subtle title *Getting the Shaft: The Radioactive Waste Controversy in Manitoba* (Robbins 1984). Toward the end of phase one, AECL's public credibility would reach historic lows, with one NGO declaring that AECL was "completely out of touch with Canadian values," "discredited" because of questionable foreign reactor sales practices, and "unfit" to be the implementing agency because it "has so abused the public trust."[70]

In an image combining interpersonal metaphors and references to institutional behaviour, one NGO thus succinctly stated that industry groups do not have "clean hands."[71] Not only did critics fear the consequences of matters being left in such hands after the inquiry was completed, but some also felt "dirty" participating in a process that may actually be co-option: "There are certain of us here today that feel we're co-opting ourselves just by participating in the process to begin with, that we're somehow legitimizing a concept that we don't agree with ... I feel completely co-opted by being here, and that I am adding an air of legitimacy to an otherwise ... bizarre ... ludicrous thing."[72]

Thus, while neither AECL nor Ontario Hydro was considered trustworthy, the specific problem was the "housing" of science rather than scientists per se. One respondent questioned several technical assertions but reminded the audience that he had "a great deal of respect for science, and I trust you do too."[73] Others noted that society needs experts, even if those experts do not always frame their claims properly.[74] Critics thus valued expertise, as when one NGO humorously argued that "we don't allow Homer Simpson to run our nuclear generating stations ... We insist on people of some qualifications and training."[75] The problem concerned the relationship between expertise and organizational interests. NGOs considered two options: skills existed if they could be freed from "the corporate umbrella," and experts were "tainted for life by their association with the tainted organization."[76] Managing relations of trust was thus a common problem. Technical and government groups often argued for sympathy as a means to develop trust. The AECB noted the difficult transition from a "purely scientific and technical organization to one which is much more sensitive to these broader issues."[77] The Canadian Nuclear Society (CNS) objected to references to "the public feels this or the public feels that," when its members were also part of the public.[78] AECL asked for a more sympathetic appreciation that it was "working hard" and wanted "nice things."[79] Even critics claimed that Canadian-based organizations were at least "accountable to the Canadian public."[80]

Yet *individual* scientists or policy makers were rarely at issue. Lack of trust in broader organizational and institutional actors was the issue. In such a situation, critics elevated the desired level of certainty because untrustworthy groups were considered unlikely to make the right decision in a context of ambiguity and interpretive flexibility. One critic voiced a common refrain, arguing that "a permanent solution demands failsafe guarantees. The experts cannot give this. Not one of the experts will be responsible for failure ... AECL's proposal is based on self-interest ... The wider view ...

[is] dependent on a measure of disinterest."[81] Rather than an idealized image of science characterized by absolute certainty and zero ambiguity, most critics acknowledged inevitable uncertainty. This became evident in phase two of the inquiry, when critics complained that AECL's disposal concept became increasingly vague as new illustrations of the concept were introduced. AECL thus introduced a new reference case: from the original titanium canisters emplaced in boreholes in the floors of disposal rooms (buried in low-permeability crystalline rock) to copper canisters emplaced directly in disposal rooms (buried in moderately permeable crystalline rock) (Durant 2009b).

Critics found unacceptable not flexibility and uncertainty per se but that flexibility wielded by untrustworthy organizations amounted to granting political discretion to a "despot."[82] Moreover, flexibility in the concept, under such circumstances, substituted a "vague learning process" for what should be "a test within set, formal terms."[83] The Scientific Review Group (SRG) sided with the public in this matter; although approving of the new information, they argued that "the public wants some brackets, some frame within which safety of any example is assured."[84] For critics, establishing the interests of AECL and Ontario Hydro, by drawing connections with the past, was a means to contextualize likely behaviour in the future. NGOs thus referred to "the lack of honesty, openness and fairness which has characterized the nuclear debate at the federal level for as long as I can remember."[85] NGOs claimed that the "credibility of [the] proponent [AECL]" was low due to their "own very negative experience with the proponent."[86]

Although technical and policy elites attempted to shift matters to a future political terrain, critics tried to shift matters to more local and immediate levels. They perceived that any future terrain may lack broad participation. Consistent with the view of Giddens (1990) that dissenting voices within modernity seek to reject the imposition of hegemonic meanings, by re-establishing local control, critics placed the practices of AECL and Ontario Hydro within a local context of community action. These practices were represented in terms consonant with evaluations of interpersonal contact: "So what is wrong with the technical stuff, the scientific stuff? It's probably very good. I do not know. But is it credible, because the interests of the agency ha[ve] been mixed with the scientific endeavour, and that is very hard when you are going to go to the public and say, 'Believe me, believe my science.'"[87]

Conflicting Social Relations

Clearly, the public inquiry inherited a problem concerning lack of trust, about which it could do little. The distrust was also mutual. In phase one, nuclear industry workers claimed that anti-nuclear activists had a "vested interest" (raising public concern and financial support) in critiquing the disposal concept.[88] In phase two, critics addressed such claims, distinguishing the "true scientist" from the "technocrat ... [who] will say what he is paid to say."[89]

AECL's R-Public (Greber, French, and Hillier 1994, 64) had attempted to bridge the trust gap, citing risk perception literature (Otway 1992; Slovic 1992) as evidence that it had addressed socially valued dimensions (e.g., voluntarism). Yet both Otway and Slovic cautioned against taking for granted the competence and trustworthiness of managing institutions. The narrow mandate of the inquiry threatened to translate concept approval into both default approval for nuclear expansion and the installation of already distrusted organizations as implementers. That the merits of having AECL and Ontario Hydro as implementers were being taken for granted was clear in phase two, when the public learned that a *Policy Framework for Radioactive Waste* had been formulated by industry-government discussion and released in July 1996 (halfway through the public inquiry). This framework presumed, citing the "polluter pays" principle, that waste owners and producers ought to form an implementing agency (Durant 2009a; Durant and Stanley, this volume).

The meaning of the disposal concept was thus tied up with competing conceptions of reasonable political action. Concept supporters emphasized the unavoidability of dealing with nuclear waste regardless of energy policy futures. Yet this pragmatic will-to-action implied political sterility for critics, who denied that concept approval carried no implications for energy policy. For supporters of the concept, doing nothing constituted an irresponsible ignoring of potential hazards. For critics, the industry was already irresponsible, having developed commercial nuclear power prior to demonstrating a solution to waste disposal. Given that many supporters of the concept admitted that approval strengthened the nuclear option (AECL and Ontario Hydro denied the relation), critics justifiably focused on the end game (nuclear expansion). Hence, they treated the waste disposal concept as a vehicle for opening up policy options for nuclear power supporters. In turn, critics demanded a nuclear phase out (Stevenson 2003, 91-92).

Yet this dispute over energy futures was, for critics, conducted on a playing field lacking any semblance of balance. The social experience of AECB and NRCan consisted of taking for granted a continued bureaucratic presence in policy making. AECL had already argued in its EIS that those who are currently responsible for, and owners of, used nuclear fuel should also be responsible for implementing the disposal concept (1994a, 344-45). Ontario Hydro agreed that "waste producers are accountable and are responsible ... for the cradle to grave management of all of our wastes."[90] Hence, AECL and Ontario Hydro also took for granted their continued policy presence. NRCan confirmed that the government supported AECL and Ontario Hydro's central role, early in the inquiry, obliquely referring to the *Policy Framework for Radioactive Waste* when suggesting the need to "start to determine preferred positions on implementation issues."[91]

Such comfortable political knowledge on behalf of supporters of the concept allowed the broad framing of opposition claims to be marginalized, as outside the mandate. Dominant groups could afford to combat claims alone, for they were comfortable in the knowledge that those actually making them were unlikely to sit at decision-making tables. For those accustomed to exclusion from decision making, the pressing political task was how to avoid an inquiry, arguably a "temporary creature," being used to hand the nuclear industry a concept deemed "flawless."[92] Politics per se was not an issue; indeed, critics welcomed the "return to the political arena [of] decisions of an essentially political nature."[93] The salient issue for critics was the presence of an undemocratic level of *already made* political decisions. Political decisions that seemed to be "irrevocable,"[94] or tacit political commitments that might "prejudge" technical assessments and options,[95] were the kind of political positions that critics wished to marginalize.

With a view to marginalizing claims denying that approval of the concept carried energy policy implications, and because the presumption that waste owners needed to be the central actors in possible concept implementation was so engrained, critics collapsed judgments about the implications of the deep geological disposal concept into the decision on the concept itself. This hermeneutics of suspicion extended to their own role in the inquiry process, for instance whether industry and policy elites saw "public opinion as an acceptability problem to be solved."[96] If supporters of the concept were disingenuous now, in other words, then this was cause to revisit the presumption that those actors ought to be central in any implementation. A sense of the frustration critics felt can be gauged in the following: "In some ways we are being used [by the] proponent to understand more about how they could soften the acceptability of a site or this concept, [and for the proponent to figure out how] to come into a community and sociologically [site a repository] in a correct way, that they will know then how to establish trust ... So in some ways I feel like I'm doing the job that I don't want to be doing."[97]

Conclusion

As Wynne has argued, failure to focus on the "social foundations of risk framing by expert institutions only provokes a greater sense of denigration and skepticism of authority on the part of the public" (1992, 278). Even though AECL admitted that the disposal project possessed an "unavoidable degree of uncertainty,"[98] a similar degree of reflexivity about political matters was not demonstrated.

Technical and governance groups underestimated both the extent to which democratic expectations have grown, producing "more 'critical citizens' or perhaps 'disenchanted democrats'" (Norris 1999, 27), and the ability of public groups to subject authority figures to "more searching scrutiny than they once were" (Inglehart 1999, 236). This underestimating of critics manifested, among supporters of the deep geological disposal concept, as the tacit presumption that decisions about waste disposal could still be made despite procedural restrictions on what the concept actually meant. Thus, supporters of the disposal concept constructed it as limited to a solution to a technical problem and not as implying anything beyond that. Critics responded by asserting their right to participate in deciding what the concept meant and thus in deciding what the salient issues and meanings in contention actually were. Critics constructed the deep geological disposal concept as tacitly privileging nuclear expansion plans and regarded the combination of specific commitments to nuclear expansion and general institutional inertia within policy making as democratically unacceptable.

The lesson to be drawn from the public inquiry is that public policy and technical research about nuclear waste management will lack credibility to the extent that discussion about salient issues and meanings is restricted, the science is conceived of as solely about problem solving, and the politics deemed relevant is constructed as only about isolated actions. Habermas long ago (1973) derided this conception of politics as "decisionism," which Wynne has defined as "a model in which policy and political processes are conceptualized exclusively as a series of completely unrelated specific decisions, each one of which has no interaction with any other" (2003, 410). Unfortunately, such decisionism continues to define nuclear waste management in Canada. The Seaborn panel concluded that the disposal concept lacked broad social support and recommended that an arm's-length agency be formed to assess and evaluate waste management options from the perspective of social safety (CEAA 1998). Ostensibly, the NWMO, though it is comprised of waste owners and thus far from arm's length, claims to have used a social safety perspective in recommending APM in November 2005 (2005c, 154-55). APM leaves open to future political bargaining which waste management option will be decided on (e.g., various forms of storage, permanent disposal, extent of monitoring) and what kind of implementation schedule will be observed (Durant 2009b).

For the purposes of discussion here, I wish to point to several features of the NWMO and APM that reinvent the deficiencies characteristic of the public inquiry. The first is that the NWMO recapitulates both the public inquiry restriction on discussion and the shallow politics of treating APM as unrelated to other possible decisions. Although the Nuclear Fuel Waste Act of 2002 does not mandate the NWMO to discuss broader matters, it does not prohibit it from doing so. The NWMO thus discursively manages the public-private and science-politics boundaries that it straddles (Durant 2006), drawing on its own discretion to situate itself at the intersection of a host of convenient separations (Durant 2007a). Thus, APM is separated from energy policy implications: "Our study process and evaluation of options were intended neither to promote nor penalize Canada's decisions regarding the future of nuclear power" (NWMO 2005c, 20). Indeed, the NWMO separates energy policy discussions entirely from APM-related matters: "Those future decisions should be the subject of their own assessment and public process" (20). The NWMO also separates the finding of a technical method from how that method is brought before the public: the "most profound challenge does not lie solely in finding an appropriate technical method, but also in the manner in which the management approach is implemented" (162).

This is decisionism par excellence. It allows the NWMO to steadily refine technical methods and present each configuration for consideration by the public, in isolation of considerations of *anyone's* institutional commitments or possible energy policy implications. The NWMO is not unique in this kind of political action, with other nuclear waste management organizations (including in Sweden, the United Kingdom, and the United States) practising similar kinds of decisionist styles (Durant 2007a). The suicidal politics involved in accepting such decisionism is suggested by the way in which the federal government accepted APM as an "initiative vital to the future of nuclear energy in Canada" and a part of "steps toward a safe, long-term plan for nuclear power in Canada" (NRCan 2007). The appropriate political analogy involved is that, within the NWMO scheme, the public plays the role of an isolated voter. When a majority vote pertains, a technical method is accepted. This is the analogue of the public inquiry situation where it was assumed that the public could "vote" on the disposal concept in isolation of, for instance, both the human purposes driving the proposal and the associated policy implications of acceptance.

Hence, I end the chapter here with some cautions. One is that restrictions on discourse can be instruments of power (see Bourdieu 1991). Another is that public consultation exercises can fall far short of contemporary aspirations that participation enables "the soft – orientations, hopes, ideas, and people's interests" – to triumph over "the hard – the organizations, the established, the powerful, and the armed" (Beck 1992, 117). But on a different note, those who value deliberative democracy, which if their rhetoric is sincere apparently includes the NWMO, need to remain committed to the value and potential efficacy of public participation in conceptualizing what the waste management issue actually is and which decision to make about it. A properly framed discussion, minus artificial separations and senseless restrictions, remains the tonic suggested by deliberative democrats. In part, this is because "real argument makes moral insight possible" (Habermas 1990, 15). Unfortunately, what we have at present in the shape of APM is not real argument but effectively a nuclear waste management implementation *concept*, leaving me with an uneasy sense of déjà vu.

Notes

1 Blair Seaborn (panel chair), testimony of 11 March 1996, *Hearings Transcripts* [vol. 1], 7, in CEAA (1997). This chapter draws on verbal testimony to the Seaborn panel, supplemented by written submissions to it. To facilitate individual identity, group affiliation, and source of evidence, I use endnotes of the following form. Direct verbal testimony, as found in the hearings transcripts (CEAA 1997), is cited as name (group affiliation), testimony of [date of testimony], *Hearings Transcripts* [abbreviated as *HT*] [volume number], page number(s). Quotations from the written submissions (CEAA 1996) are cited as name (group affiliation), *Written Submissions* [abbreviated as *WS*] [volume number], register number, page number(s).

2 Dr. Peter Brown (NRCan), testimony of 11 March 1996, *HT* [vol. 1], 30.

3 Gordon Edwards (Canadian Council for Nuclear Responsibility), testimony of 11 March 1996, *HT* [vol. 1], 36-37.

4 Dr. Peter Brown (NRCan), testimony of 11 March 1996, *HT* [vol. 1], 38-39.

5 Brennain Lloyd (Northwatch), testimony of 11 March 1996, *HT* [vol. 1], 225.

6 Blair Seaborn (panel chair), testimony of 11 March 1996, *HT* [vol. 1], 228.

7 J.A.L. Robertson (concerned citizen), *WS* [vol. 4], PHPUB.004, 8-9. Robertson was a long-time AECL employee.

8 Ken Nash (Ontario Hydro), testimony of 11 March 1996, *HT* [vol. 1], 56-57.

9 David Smith (Canadian Academy of Engineering and Royal Society of Canada), testimony of 25 March 1996, *HT* [vol. 6], 129-30.

10 Ken Bragg (AECB), testimony of 11 March 1996, *HT* [vol. 1], 147.

11 Jerry Cuttler (Canadian Nuclear Society), *WS* [vol. 1], PHPUB.22, 3.

12 David Shier (president of the Canadian Nuclear Workers Council), testimony of 25 March 1996, *HT* [vol. 6], 153.

13 J.A.L. Robertson (concerned citizen), testimony of 2 May 1996, *HT* [vol. 13], 112.

14 Ian Wilson (CNA), testimony of 29 March 1996 (in response to Brennain Lloyd [Northwatch]), *HT* [vol. 10], 60-63.

15 Dr. Peter Brown (NRCan), testimony of 11 March 1996, *HT* [vol. 1], 38-39.

16 Dr. Peter Brown (NRCan), testimony of 12 March 1996, *HT* [vol. 2], 9.

17 Peter Prebble (Saskatchewan Environmental Society), testimony of 29 March 1996, *HT* [vol. 10], 194-96; Dave Plummer and Ann Lindsay (Concerned Citizens of Manitoba), testimony of 29 April 1996, *HT* [vol. 11], 37-41.

18 Ian Wilson (CNA), testimony of 29 March 1996, *HT* [vol. 10], 205.

19 Norm Rubin (Energy Probe), testimony of 29 March 1996, *HT* [vol. 10], 211-12.

20 Graham Latonas (Chem-Security Limited, Alberta), testimony of 25 March 1996, *HT* [vol. 6], 174.

21 Avro Ranni (private citizen), testimony of 11 March 1996, *HT* [vol. 1], 55-56.

22 Dave Plummer and Ann Lindsay (Concerned Citizens of Manitoba), *WS* [vol. 4], PHPUB.153, 5.

23 Peter Prebble (Saskatchewan Environmental Society), testimony of 29 March 1996, *HT* [vol. 10], 35.

24 Gordan Edwards (Canadian Council for Nuclear Responsibility), testimony of 15 March 1996, *HT* [vol. 5], 50-52.

25 Brennain Lloyd (Northwatch), testimony of 29 March 1996 (questioning Ian Wilson [CNA]), *HT* [vol. 10], 60-63.

26 Irene Koch (Nuclear Awareness Project), testimony of 26 March 1996, *HT* [vol. 7], 141.

27 Tom Lawson (concerned citizen), testimony of 27 March 1996, *HT* [vol. 8], 70.

28 Glenn Kukkee (concerned citizen), testimony of 29 April 1996, *HT* [vol. 11], 84.

29 Malvina Iron (Alberta Indigenous Women Environmental Foundation), testimony of 28 March 1996, *HT* [vol. 9], 165.

30 Sue Allen (concerned citizen), testimony of 25 March 1996, *HT* [vol. 6], 154.

31 Walter Robbins (CONSUN), testimony of 28 March 1996, *HT* [vol. 9], 96.

32 Evelyn McNenly (Town of Massey), testimony of 30 April 1996, *HT* [vol. 12], 30-31.

33 Deputy Grand Chief Davey (Nishnawbe-Aski Nation), testimony of 11 March 1996, *HT* [vol. 1], 243.

34　Vice Chief Allan Adams (Federation of Saskatchewan Indians), testimony of 11 March 1996, *HT* [vol. 1], 259-60.

35　Andrew Orken (speaking on behalf of Grand Chief Fontaine and on behalf of the Quebec and Labrador region of the Assembly of First Nations and the Grand Council of the Crees [of Quebec]), testimony of 11 March 1996, *HT* [vol. 1], 359.

36　Fred Bianchi and Rita O'Sullivan (Aboriginal Rights Coalition), testimony of 29 April 1996, *HT* [vol. 11], 16-17.

37　Stella Swanson (Golder Associates), testimony of 15 March 1996, *HT* [vol. 5], 135. Swanson was also a member of the Scientific Review Group (SRG).

38　John Davey (Ontario Association for Environmental Ethics), testimony of 11 March 1996, *HT* [vol. 1], 132.

39　Brennain Lloyd (Northwatch), testimony of 30 April 1996, *HT* [Vol. 12], 51 and 45-60.

40　Betsy Carr (concerned citizen), testimony of 15 March 1996, *HT* [vol. 15], 66-67.

41　Dr. Robert Paehlke (Trent University [invited speaker]), testimony of 29 March 1996, *HT* [vol. 10], 122.

42　Ken Bragg (AECB), testimony of 13 March 1996, *HT* [vol. 3], 26, 18-19.

43　Dr. Mary Measures and Kate Maloney (AECB), testimony of 3 May 1996, *HT* [vol. 14], 59.

44　Denis Hall (Deep River Low-Level Radioactive Wastes Siting Task Force), testimony of 25 March 1996, *HT* [vol. 6], 64.

45　Dr. Vera Lafferty (Deep River Low-Level Radioactive Wastes Siting Task Force), testimony of 27 March 1996, *HT* [vol. 8], 15-16, 21.

46　Robert W. Pollock (Low-level Radioactive Waste Management Office, NRCan), testimony of 25 March 1996, *HT* [vol. 6], 25.

47　Jean-Pierre Olivier (Nuclear Energy Agency, OECD), testimony of 25 March 1996, *HT* [vol. 6], 77-97.

48　Dave Thompson and Donna Oates (Insights and Solutions), testimony of 2 May 1996, *HT* [vol. 13], 69.

49　Dr. Paul Tamblyn (Acton High Scool), testimony of 12 March 1996, *HT* [vol. 2], 137.

50　J.A.L. Robertson (concerned citizen), testimony of 2 May 1996, *HT* [vol. 13], 112-13. Again, it is worth noting here that Robertson was a long-time AECL employee and has authored many strident attacks on anti-nuclear positions.

51　Dennis Baker (concerned citizen), testimony of 27 March 1996, *HT* [vol. 8], 89.

52　Dr. Bill Leiss (Philosophy, Queen's University), testimony of 15 March 1996, *HT* [Vol. 5], 94-118.

53　Ella de Quehen (Northumberland Environment Protection), testimony of 13 March 1996, *HT* [vol. 3], 157.

54　Marion Penna (Inter-Church Uranium Committee), testimony of 26 March 1996, *HT* [vol. 7], 97.

55　Norm Rubin (Energy Probe), testimony of 13 March 1996, *HT* [vol. 3], 80-81.

56　Norm Rubin (Energy Probe), testimony of 11 March 1996, *HT* [vol. 1], 77-78.

57　Dr. Peter Brown (NRCan), testimony of 11 March 1996, *HT* [vol. 1], 35-40.

58　Ken Bragg (AECB), testimony of 13 March 1996, *HT* [vol. 3], 11.

59　Ken Nash (Ontario Hydro), testimony of 11 March 1996, *HT* [vol. 1], 44.

60　Dr. Ken Dormuth (AECL), testimony of 11 March 1996, *HT* [vol. 1], 72-76.

61　Gordon Edwards (Canadian Coalition for Nuclear Responsibility), testimony of 15 March 1996, *HT* [vol. 5], 41.

62　Ella de Quehen (Northumberland Environment Protection), testimony of 13 March 1996, *HT* [vol. 3], 145-64.

63　Patricia Lawson (Environmental Protection Group) and Marc Chenier (Campagne contre l'expansion du nucleaire), testimony of 14 March 1996, *HT* [vol. 4], 77-85, 170-85.

64　Irene Kock (Nuclear Energy Awareness), testimony of 26 March 1996, *HT* [vol. 7], 136-45.

65　Fred Long (Ontario Hydro), testimony of 29 March 1996, *HT* [vol. 10], 8-26.

66　Tom Lawson (Port Hope Citizens for Responsible Management of Radioactive Waste), testimony of 27 March 1996, *HT* [vol. 8], 71-72.

67　Malvina Iron (Malvern Indigenous Women Environmental Foundation), testimony of 28 March 1996, *HT* [vol. 9], 155-69.

68 Fred Bianchi and Rita O'Sullivan (Aboriginal Rights Coalition), testimony of 29 April 1996, *HT* [vol. 11], 12-28.
69 David Plummer and Ann Lindsay (Concerned Citizens of Manitoba), testimony of 11 March 1996, *HT* [vol. 1], 1-50.
70 Peter Prebble (Saskatchewan Environmental Society), testimony of 29 March 1996, *HT* [vol. 10], 32-37, 191-94.
71 Lloyd Greenspoon (Algoma-Manitoulin Nuclear Awareness), testimony of 30 April 1996, *HT* [vol. 12], 16.
72 Julie Dingwell (People against Lepreau 2), testimony of 27 March 1996, *HT* [vol. 8], 164; testimony of 28 March 1996, *HT* [vol. 9], 57-58.
73 Bill Crowley (Port Hope), testimony of 25 March 1996, *HT* [vol. 6], 121.
74 Dr. Bill Leiss (Queen's University, Department of Philosophy), testimony of 15 March 1996, *HT* [vol. 5], 107.
75 Norm Rubin (Energy Probe), testimony of 10 June 1996, *HT* [vol. 15], 144.
76 Norm Rubin (Energy Probe), testimony of 29 March 1996, *HT* [vol. 10], 211-12.
77 Ken Bragg (AECB), testimony of 13 March 1996, *HT* [vol. 3], 43.
78 Ken Smith (CNS), testimony of 14 March 1996, *HT* [vol. 4], 186.
79 Colin Allen (AECL), testimony of 12 March 1996, *HT* [vol. 2], 327-29.
80 Walter Saveland (concerned citizen), testimony of 3 May 1996, *HT* [vol. 14], 10.
81 Patricia Lawson (Environmental Protection Group), testimony of 14 March 1996, *HT* [vol. 4], 70, 84.
82 Norm Rubin (Energy Probe), testimony of 20 June 1996, *HT* [vol. 23], 174-76.
83 Ella de Quehen (Northumberland Environmental Protection), testimony of 19 November 1996, *HT* [vol. 28], 157-58.
84 Stella Swanson (SRG), testimony of 19 June 1996, *HT* [vol. 22], 138.
85 Gordon Edwards (Canadian Coalition for Nuclear Responsibility), testimony of 11 March 1996, *HT* [vol. 1], 268.
86 Brennain Lloyd (Northwatch), testimony of 11 March 1996, *HT* [vol. 1], 219.
87 Maria Paez-Victor (Voice of Women), testimony of 15 March 1996, *HT* [vol. 5], 207.
88 David Shier (president of the Canadian Nuclear Workers Council), testimony of 25 March 1996, *HT* [vol. 6], 142.
89 Tom Lawson (Port Hope Citizens for Responsible Management of Radioactive Waste), testimony of 11 June 1996, *HT* [vol. 16], 197.
90 Ken Nash (Ontario Hydro), testimony of 11 March 1996, *HT* [vol. 1], 44-45.
91 Dr. Peter Brown (NRCan), testimony of 11 March 1996, *HT* [vol. 1], 30.
92 Gordon Edwards (Canadian Council for Nuclear Responsibility), testimony of 11 March 1996, *HT* [vol. 1], 282.
93 Walter Saveland (concerned citizen), testimony of 3 May 1996, *HT* [vol. 14], 131-32.
94 Dave Plummer (Concerned Citizens of Manitoba), testimony of 14 March 1996, *HT* [vol. 4], 49-50.
95 Norm Rubin (Energy Probe), testimony of 11 March 1996, *HT* [vol. 1], 34.
96 Chaitanya Kalevar (concerned citizen), testimony of 14 March 1996, *HT* [vol. 4], 166.
97 Marion Penna (Inter-Church Uranium Committee), testimony of 25 March 1996, *HT* [vol. 6], 235-36.
98 Mary Grebber (AECL), testimony of 12 March 1996, *HT* [vol. 2], 266.

6

The Darker Side of Deliberative Democracy: The Canadian Nuclear Waste Management Organization's National Consultation Process
Genevieve Fuji Johnson

On the surface, the Nuclear Waste Management Organization's (NWMO) national consultation process was an impressive attempt to realize principles associated with the ideal of deliberative democratic citizen engagement (see NWMO 2005c). It appears to have been one of Canada's most serious endeavours to implement aspects of deliberative democratic decision making in a public policy process. Its apparent aims would be laudable given the ethical importance of the deliberative democratic ideal for policy that is not only publicly binding but also, perhaps more importantly, socially and environmentally risky. These aims would be laudable, moreover, given the practical implications of deliberative democracy for policy formulation and implementation in areas traditionally dominated by government and industry elites, such as nuclear energy and nuclear waste management.

This chapter takes a critical perspective on the NWMO's national consultation. Based on a qualitative analysis of the NWMO's discussion documents, submissions to its consultation process, and interviews with selected actors, I argue that the NWMO's process was essentially non-deliberative, reinforcing inequalities between a dominant coalition of the nuclear energy industry and a critical coalition of concerned religious and environmental organizations and Aboriginal nations. Perhaps more seriously, the process served in yoking the industry's preformed policy preference for deep geological disposal with a questionable degree of legitimacy. This critical examination reveals how certain attempts to realize deliberative democratic decision making can lend themselves to purposes antithetical to the deliberative ideal. In certain contexts, such endeavours can contribute to entrenching power relations between policy coalitions and provide a veneer of legitimacy to pre-established policy preferences.

Deliberative democracy is an ideal, premised on the axiom of the fundamental moral equality of persons. As an ideal, it prescribes upholding this axiom vis-à-vis the binding nature of public policy. It counsels including in policy processes all persons who will be affected by those processes' outputs.

Specifically, it instructs policy makers to ensure that all affected persons (or even all potentially affected persons) exercise their decisional agency in these processes.[1] It instructs them to ensure that affected or potentially affected persons participate in these processes under conditions of freedom and equality.[2] Each participant should be free to express his or her perspectives and to raise and respond to questions.[3] Each should have access to the information and information technologies necessary to deliberate as an equal. Moreover, each should put forth reasons in support of his or her position on policy options that are generally comprehensible and acceptable to all.[4] Public reasoning involves participants willing to reflect on broader shared interests and to advance arguments that promote these interests. Ideally, in light of compelling arguments concerning broader shared interests, participants reconsider and modify their positions in order to find the common ground necessary for justifiable or provisionally justifiable agreement on the policy.[5]

The ideal of deliberative democracy may give rise to the most ethically defensible way of making binding decisions, especially in areas associated with high-magnitude risk and great uncertainty (Johnson 2008). In such areas, science-based decision making is often ethically insufficient given the high social, environmental, and economic stakes and a corresponding lack of information. Standard approaches to risk assessment generally do not incorporate the range of values at stake and cannot adequately address gaps in knowledge about future impacts. Deliberative democracy can highlight ways of complementing positivist decision making to provide a justifiable or provisionally justifiable moral basis for policy. But its implications are far reaching, extending to the motivation of policy stakeholders and deliberative actors, the design of policy processes, and pre-existing power structures that often exist between coalitions of traditional and coalitions of non-traditional policy actors. These implications are particularly pronounced in policy areas historically dominated by elites from the realms of government, industry, and science. Dominant coalitions of industry and government tend to set the policy agenda and advance their own policy preferences to the exclusion of other perspectives and interests. Dominant coalitions tend to marginalize coalitions of environmental, religious, and Aboriginal organizations, which thus tend to play a more reactive role in the policy area. The NWMO's recent consultation process may be seen as a noteworthy attempt ethically to justify an important policy decision and address the historical power inequality between two coalitions in the area of Canadian nuclear waste management policy. In this chapter, I challenge this perspective.

The NWMO's National Consultation Process

As directed by the Nuclear Fuel Waste Act, Canada's three nuclear energy corporations (Ontario Power Generation [OPG], Hydro-Québec, and New

Brunswick Power [NBP]) and Atomic Energy Canada Limited (AECL) established the NWMO in 2002 (Parliament of Canada 2002). In accordance with this legislation, the nuclear energy corporations and AECL set up a segregated fund to finance the activities of the NWMO. According to the act, these activities must include assessing at least three options for Canada's growing stockpile of irradiated nuclear fuel bundles (e.g., on-site storage, centralized storage, and deep geological disposal), recommending to the federal government an option, and implementing the option ultimately chosen by the federal government.[6]

The NWMO may appear to be distinct from the Canadian nuclear energy industry. However, the organization's staff and board have roots in the Canadian nuclear energy industry. Throughout the consultation process, the NWMO's Board of Directors included only representatives of the nuclear energy corporations and AECL. A number of the NWMO's staff, moreover, came from OPG. Many members of the NWMO's assessment team are or were employees of OPG, the Canadian Nuclear Safety Commission (CNSC), the US Department of Energy, Nuclear Division, or the NWMO. The close relationship between the industry and the organization is reinforced by a weak oversight arrangement. The board appointed its own Advisory Committee to make yearly reports on the organization's activities. Although the organization reports to Parliament on a yearly basis, is to pay penalties where it violates the Nuclear Fuel Waste Act, and is subject to the regulation of the CNSC once it becomes licensed to implement a waste management system, it is not subject to the auditor general or the Access to Information law. The NWMO – an industry-based organization that has been formed, funded, and staffed by the nuclear energy corporations and AECL – has a significant degree of discretion in fulfilling its legislative obligations.

Soon after its establishment, the NWMO launched a three-year national consultation process toward the end of assessing waste management options and developing "collaboratively" a management approach that is socially acceptable, technically sound, environmentally responsible, and economically feasible" (NWMO 2005c, 17). Designed and directed by the NWMO, the process included at least twenty sets of dialogues organized over four phases. During each phase, staff at the NWMO focused dialogue sessions on a key decision. NWMO staff commissioned firms and organizations specializing in deliberative approaches to decision making. They also asked academics and community leaders to run certain dialogues. It is important to point out the distinction between the NWMO and these firms, organizations, and groups of academics and community leaders. The latter played a primarily "third-party" role in the process, providing professional services or specialized advice to the organization. The former was in large part responsible for framing, organizing, and funding the process as a whole, identifying and recruiting many – although not all – of the dialogue participants, establishing and phrasing the general dialogue questions, interpreting the results of each dialogue, and, ultimately, formulating a policy recommendation to the federal government. The NWMO, in other words, played a crucial role in the consultation process.

After each dialogue, the firm, organization, or group of academics or community leaders would report its findings and conclusions to the NWMO. On the basis of its interpretation of these reports, the NWMO would write and publish a discussion document encapsulating its decision for the phase. The document would, in turn, be the focus for the following phase, which would seek public validation of the previous decision and direction on the next decision to be taken in the subsequent phase. Throughout the entire process, the NWMO's website offered a public platform not only for dialogue reports, discussion documents, and background papers but also for further public submissions.

The designers of the NWMO's national consultation process claim to have sought to realize principles associated with deliberative democracy. For instance, they claim to have premised the process on the understanding that the insights of all those affected by the policy, not just those of industry representatives and technical and scientific experts, are important and should be incorporated into their recommendation to the federal government (Dowdeswell 2005; NWMO 2005b, 29-40; Staff, NWMO 2005). To be sure, many of the dialogues were based on procedural principles calling on a range of participants to explore and exchange informed reasons and arguments in a respectful manner (see, e.g., DPRA Canada 2004b; Hardy, Stevenson, and Associates 2005a, 2005b; Sigurdson and Stuart 2003; and Watling et al. 2004). Dialogue participants, moreover, had a certain equality of access to information. They had access to the organization's website, which housed more than sixty peer-reviewed papers from a variety of disciplines, methodologies, and ideologies (NWMO 2005b, 262-66). The process also involved more than 120 public information sessions held across the country (31) and several e-dialogues, which contributed to the knowledge base of participants (33). Based on this process, the NWMO recommended to the federal government an Adaptive Phased Management (APM) system composed of continued on-site storage, centralized storage, and deep geological disposal and complemented by continuous monitoring, provisions for retrievability, and procedures for citizen engagement (NWMO 2005c, 23-30). On the surface, the NWMO's consultation process was informed by principles of inclusion, procedural equality, informational equality, and public reasoning. A deeper examination reveals questions about the extent to which these principles obtained.

Critical Questions

The first questions that arise concern the inclusive nature of the NWMO's process. The organization's process engaged individuals broadly reflective of the Canadian adult population, individuals with diverse experiences and expertise in various aspects of nuclear waste management, and individuals, organizations, and Aboriginal nations with a publicly articulated interest in nuclear waste management policy (NWMO 2005b, 267-71; Staff, NWMO 2005). The dialogues of the Canadian Policy Research Networks (CPRN) were particularly noteworthy. The NWMO commissioned the CPRN to facilitate exchanges among a statistically representative sample of the Canadian adult population (Watling et al. 2004). These dialogues brought together 462 individuals with views on nuclear energy consistent with those of the broader population. The NWMO also commissioned firms to run stakeholder roundtables. For example, DPRA Canada (2004b) ran a roundtable with a variety of individuals and organizations with an interest in Canadian nuclear waste management policy. To provide assurance that the group of participants would reflect the array of interests at stake, the NWMO with the assistance of DPRA Canada identified prospective participants in terms of interest categories (e.g., "Local/Municipal," "Education/Academic," "Cultural/Faith," "Industry/Economic," "Labour," "Consumer," "Environment," "Health," and "Youth"). A year later Hardy, Stevenson, and Associates (2005a) reconvened many of these stakeholders. Where they were unable to include the same participants, they employed this template to ensure a general reflectiveness of their participants. In the summer of 2005, Stratos (2005) employed the same template for its stakeholder dialogues. The NWMO retained a say in determining who would participate in these stakeholder dialogues.[7]

The NWMO also organized expert workshops and roundtables. In September 2003, it appointed an ethics roundtable that included individuals from a range of professional and academic backgrounds (2004c). The roundtable included not only academics in the disciplines of philosophy, medicine, business, and political science but also community leaders with perspectives informed by Aboriginal traditional knowledge and Christianity. The NWMO also commissioned a workshop with experts in the technical aspects of nuclear waste management (Shoesmith and Shemilt 2003); another with experts from business, industry, non-governmental organizations (NGOs), and the public sector (Coleman, Bright Associates, and Patterson Consulting 2003); and another with Aboriginal people, nuclear workers, environmentalists, academics, and members of religious organizations (Global Business Network [GBN] 2003). Again, the NWMO played a direct role in determining who would participate in these meetings.[8]

In addition, the NWMO sought input from the Assembly of First Nations (AFN), Congress of Aboriginal Peoples (CAP), Inuit Tapiriit Kanatami (ITK), Métis National Council, Native Women's Association of Canada (NWAC), and Pauktuutit Inuit Women's Association. The NWMO funded these organizations to run their own dialogues in order to improve the capacity of Aboriginal peoples to participate more effectively in nuclear waste management policy and build more effective working relationships between national Aboriginal organizations and the NWMO (2005b, 267). Other stated objectives were to understand Aboriginal perspectives on technical waste management and disposal options and to include Aboriginal ideas, insights, wisdom, and values in the NWMO's recommendation to government.

Although the NWMO's process was fairly inclusive of a range of traditional and non-traditional actors, it was exclusive of important perspectives and arguments.[9] The NWMO's approach to inclusion reveals a crucial distinction between including actors and including their perspectives and arguments. Organizers of a deliberative democratic process can invite an inclusive range of actors to participate without necessarily including the range of their perspectives. Organizers can, in other words, set the parameters for who deliberates, what is up for deliberation, and which views should be included in deliberation. Prima facie inclusion thus refers to actors but not necessarily their views or, more pointedly, their voices. When looking at the NWMO's process, it is clear that it was in this sense inclusive. The NWMO included in its process a wide range of actors; however, it excluded critical voices from its dialogues, reports, and recommendation.

In particular, the NWMO excluded important voices concerning the place of nuclear in Canadian energy policy and, by extension, the nature of the nuclear waste problem and prospects for a safe and acceptable waste management solution. According to religious, environmental, and other NGOs and independent observers, the NWMO excluded such concerns despite participants consistently raising and identifying them as crucial in the assessment of nuclear waste management solutions.[10] As stated in its submission to the NWMO in its final report, "the United Church has been concerned throughout the NWMO process by the exclusion of aspects from discussions and assessments that were repeatedly raised by participants. As a necessary first step, social acceptability must be addressed in the full context of the issues in which it is perceived by society, including the full complex of problems in the nuclear fuel cycle ... and the future role of nuclear power in Canadian energy and export policy" (United Church of Canada 2005b, 1).

Some argue that these concerns fall beyond the NWMO's legislative mandate for its consultation process. However, the Nuclear Fuel Waste Act does not explicitly state that management and disposal systems involving clarity on the total volume of waste should not be included. According to critics, without clarity on the future of nuclear energy in Canada, without clarity on plans either to expand or phase out nuclear energy, it is impossible to determine the extent of the nuclear waste management problem. It is impossible to determine the amount of nuclear fuel waste that would be contained by a storage or disposal system. It is impossible to assess any waste management or disposal system. Moreover, by excluding discussion on the status of nuclear energy in Canada, the NWMO excluded discussion of a viable waste management solution or part of one: reduction at source. Reporting on one of its dialogues, Stratos (2005, 2) writes that "Canada's environmental organizations have taken a unified position on the management of nuclear waste ... The fundamental principle [that they have] put forward is that the most effective form of waste management is controlling the production of waste at the source (i.e., not creating this waste in the first place). The focus of nuclear waste management should therefore be on managing existing waste, and reducing (eliminating) the production of additional waste."

The NWMO did not include in its set of questions that would frame the consultation process (presented in its first discussion document) questions concerning the status of nuclear in Canadian energy policy and the option of reducing or phasing out nuclear energy (2003, 8). In its second discussion document, it states that participants indicated that "the ten key questions ... capture the key issues and considerations that should be addressed" (2004e, 5). Yet, in its 2004 annual report, the NWMO acknowledges that many participants "spoke about energy policy, expressing a belief that source reduction and elimination should be the first step in any management program" (2004b, 34). It states, furthermore, that the "absence of a fully articulated plan on the future of nuclear energy is a fundamental limiting factor of the NWMO's study for those who would assess the approaches differently" (35). According to the report, these arguments "came up frequently and strongly" in the dialogues (34). In the NWMO's third discussion document, the draft study report, it is stated that "the NWMO has not examined nor is it making a judgment about the appropriate role of nuclear power generation in Canada" (2005b, 26). In light of the persistent concerns, by not examining and addressing these concerns, the NWMO made a judgment about the role of nuclear energy in Canada. In its final report, the NWMO acknowledges these criticisms and suggests that they be the subject of their "own assessment and public process" (2005c, 20). With its power to exclude, the NWMO privileged in its consultation process consideration of on-site storage, centralized storage, deep geological disposal, and, ultimately, APM over the reduction and elimination of nuclear energy.

The NWMO's consultations appear to have had other deliberative democratic merits. Indeed, organizers of the consultations made gestures toward a certain procedural and informational equality (Dowdeswell 2005; Staff, NWMO 2005). As mentioned, the NWMO commissioned firms specializing in public consultations, citizen engagement forums, and participatory decision-making processes. It instructed these firms to run dialogues that would allow for equal participation. Thus, many dialogues were premised on principles of respect for the equality of participants and their points of view (DPRA Canada 2004a, 2004b, 2004c; GBN 2003; Hardy, Stevenson, and Associates 2005a, 2005b; Sigurdson and Stuart 2003; Watling et al. 2004). Participants, furthermore, had access to the background papers, reports, and submissions posted on the NWMO's website. Where they could not access the Internet, they could request hardcopies. Some efforts were also made to bring in independent specialists to provide overviews of the technical, social, and ethical issues characterizing nuclear waste management. For instance, David Shoesmith and Les Shemilt, who organized a workshop on the technical aspects of nuclear fuel waste management, invited Phil Richardson of Enviros Consulting to make the keynote address. The CPRN, moreover, asked Blair Seaborn (chair of the Environmental Assessment of the Concept for Deep Geological Disposal) (Canadian Environmental Assessment Agency [CEAA] 1998) to review the information materials that it developed for its dialogues (Watling 2005).

There is some evidence suggesting that the NWMO's dialogues attained a degree of procedural and informational equality. For example, the CPRN reports that 91 percent of its participants agreed that there was sufficient opportunity to contribute and participate (Watling et al. 2004, 56). Mary Lou Harley (2005), who participated in a number of dialogues representing the United Church of Canada, states that "good facilitators countered the problem [of inequalities among participants] by making a point of valuing the input of each person ... They made an effort to draw back into discussion anyone who seemed intimidated and together they generally kept the discussion moving over the bumps." Dave Hardy (2001), the principal at Hardy, Stevenson, and Associates, who has worked for more than two decades in the social and ethical dimensions of Canadian nuclear waste management, also speaks of the efforts of facilitators to ensure that participants had equal opportunities.

However, there is more weighty evidence indicating that participants did not have equal opportunities in dialogues and that the consultation process was inherently hierarchical. As put by Anna Stanley, Richard Kuhn, and Brenda Murphy, various communities did

> not participate on an equal footing. For example, those from higher levels of government, the nuclear industry and other select groups tend to have more money, access to knowledge, power, and other resources to adequately understand the nuclear waste issue and to participate in the decision-making process. This leaves other communities of interest, such as the young and perhaps some place-bound communities and First Nation groups, at a comparative disadvantage. Environmental NGOs are also in a tight position, since they often have the knowledge and the desire to participate, but usually operate under severe personnel and budget constraints (Stanley, Kuhn, and Murphy 2004a).11

Aboriginal nations were particularly critical of the inherent power inequality built into the process. Although the national and regional Aboriginal organizations designed, organized, and ran their own dialogues, they expressly noted that these dialogues did not constitute formal Aboriginal consultations (AFN 2005b and 2005d; CAP 2005; ITK 2005). Aboriginal nations were not provided with the opportunity to engage directly with the Canadian government. Nor were they provided with sufficient time and funding to participate on equal footing with the NWMO. Aboriginal participants also noted that the information on the website was inaccessible (AFN 2005b; Kneen 2005; Staff #Two, AFN 2005). Many Aboriginal participants did not have access to the Internet. Moreover, the online materials, as well as the print materials, were not translated into their first languages.

Beyond questions about equal opportunities to participate, there were questions about the equal consideration of perspectives. There are many accounts of how the NWMO did not give equal consideration to perspectives critical of the place of nuclear in Canadian energy policy and the nuclear energy industry's preference for waste disposal (CAP 2005; Janes 2005; Staff #Two, AFN 2005; Stanley, Kuhn, and Murphy 2005; Trudeau Foundation and Sierra Club of Canada 2005; United Church of Canada 2004, 2005a, 2005b, 2005c). CAP notes that the "Aboriginal concerns, priorities, and values" embodied in the recommendation are merely the interpretations of the NWMO (2005, 2). As an AFN staff member notes, the NWMO drew from its dialogues with Aboriginal peoples only what would support its recommendation (Staff #Two, AFN 2005). In particular, the staff member states that the NWMO misconstrued the precept of thinking "seven generations hence," which if properly understood would rule out the nuclear generation of electricity (see also Stanley 2005). The Inuit, Métis, and First Nations argued that they did not participate in real consultation, that their Aboriginal and treaty rights were not upheld, and that their languages, cultures, and ethical principles were not respected in either the NWMO's process or its recommendation for APM (AFN 2005d; CAP 2005; Kneen 2005; Staff #Two, AFN 2005; Stanley 2005; Stanley, Kuhn, and Murphy 2005).

The NWMO seems not to have granted equal consideration to other important perspectives in their analysis of dialogue results and recommendation to the federal government for APM. From the perspectives of Nuclear Waste Watch (2004) and the United Church of Canada (2005a), the NWMO's recommendation was not based on an equal consideration of important concerns related to the future of nuclear in Canadian energy policy. Most participants in a roundtable organized by the Trudeau Foundation and the Sierra Club of Canada expressed a lack of support for the recommendation because of "concerns with the process that led to the recommendation, as well as concerns about the recommendation itself" (2005). According to the

"Proceedings, Response, and Recommendations" of the roundtable, "the study's recommendation was not socially acceptable because it failed to meet the substantive concerns raised by civil society groups about nuclear fuel waste management" (4). The report further states that the NWMO "dismissed, undervalued, and manipulated" the experience and knowledge of Aboriginal peoples and civil society groups (4).

This analysis of the NWMO's process reveals that procedural equality implies equal opportunities not merely to participate at the deliberative table but also to be taken seriously in the analysis and interpretation of deliberative results. Again, there is an important distinction between prima facie procedural equality and a deeper understanding of procedural equality. The former refers to procedural rules of dialogue that apply equally to participants. However, just because participants have equal opportunities to contribute to the dialogue, they may not necessarily have an equal say in the dialogue and ensuing recommendation. Of course, not all perspectives can or should withstand the scrutiny of public reasoning that the deliberative ideal demands. Perspectives opposing broader common interests will likely fail the test of public reason and should therefore not be admitted to a policy recommendation. The point is that deliberators should have equal opportunities to have their perspectives subject to the demands of public reasoning. Where their perspectives are informed by sound reasons and refer to a broader common interest, they should be integrated into the recommendation. This deeper kind of procedural equality implies an integration into the policy recommendation of dialogue results. Where participants validate their views by a reciprocal exchange of public reasons, they should have their views accounted for in the recommendation. Deliberative democratic decision making implies procedural equality throughout the process, from ensuring that participants have an equal voice when deliberating on policy options to ensuring that their voices transmit to the policy recommendation. Without procedural equality at the policy recommendation and implementation stages, equality at the deliberative table is largely insignificant.

Evidence concerning access to information also varies. For example, 85 percent of CPRN's participants agreed that its workbook was clear and contained relevant and useful information (Watling et al. 2004, 56). Yet only 51 percent thought that the information package sent in advance provided helpful information (56). Hardy (2005) states that participants in his dialogues had well-balanced information and had access to the NWMO's website. Berger (2005), whom the NWMO asked to comment on the consultation process, also claims that the materials were generally well balanced. A member of the NWMO's advisory council (2005) also comments on the quality of materials. Nonetheless, there were serious criticisms. Harley (2005) notes that, "at most dialogues, the presentation of the hazards of the waste was minimal and sometimes misrepresented." The NWMO's attempt to describe properly the inherent hazards of the wastes and their implications for long-term management came at the end of the study process and appeared as an appendix in its final study report.[12] The United Church of Canada, in its comments on the NWMO's draft study report, states that

> the NWMO should use the timeframe for potential health risk of "more than one million years" that is referenced (242) in the Nature of the Hazard (Appendix A) rather than "thousands of years" (e.g., 9, 12, and 66) for which there is no supporting documentation given. The material in Appendix 2 illustrates that it is over a million years before the external radiation hazard of CANDU nuclear fuel waste approaches the external radiation hazard from uranium ore and it is longer before the internal radiation potential health hazard is reduced to remain comparable to the hazard level of uranium ore (this is in the referenced material and should be illustrated in the report). Further, when using characteristics of uranium ore as the point of reference, NWMO should make it very clear that this is not safety, rather it is a level of persistent hazard for comparison. (2005a, 5)

Stanley, Kuhn, and Murphy echo these more critical views on the NWMO's information presentations (2004a, 2004b, 2005).

From this case, we see that the deliberative criterion of equal access to information is more complex than the NWMO's provisions for informational equality. It implies a certain impartiality in the informational materials provided. Impartiality is an elusive quality, especially in terms of identifying a policy problem and a range of policy solutions in a historically contentious area. In the case of nuclear waste management, where there is much social and scientific uncertainty, competing epistemologies, and normative frameworks, no single source can claim to be impartial. To approach impartiality in the provision of information about a problem as complex and divisive as nuclear waste, informational materials should derive from a range of sources reflective of the range of perspectives in the policy area. The NWMO's background papers served this purpose. Many of these papers were written by academics critical of standard decision-making tools and sensitive to social, ethical, and environmental issues in nuclear waste management. Although making these papers publicly accessible online, the NWMO did not incorporate them into its presentations of the nuclear waste problem to dialogue participants. Instead, it presented to these participants its own understanding of the waste problem. Given the NWMO's influence over the consultation in terms of framing the waste problem, setting dialogue parameters, and interpreting dialogue results, the NWMO tended to privilege its own informational materials and thus its own perspectives on the nuclear waste problem and nuclear waste management solutions in Canada.

Nonetheless, the organizers of the NWMO's consultation process sought a degree of public reasoning. Many dialogues ran according to "principles of participation," "principles of dialogue," or "principles of meaningful dialogue," all speaking to an ideal of deliberative public reasoning (DPRA Canada 2004a, 2004b, 2004c; GBN 2003; Hardy, Stevenson, and Associates 2005a, 2005b; Sigurdson and Stuart 2003; Stratos 2005). The CPRN, for example, designed these dialogues to enable participants to work through initial opinions and to reach broader and deeper understandings of nuclear waste management approaches (Watling et al. 2004). On reviewing with the facilitators the rules of dialogue, which were distinguished from those of debate, participants broke out into smaller groups in which they deliberated among themselves on the characteristics that they would most want to see in a waste management approach. Participants reconvened in plenary, where each group reported on the results of its discussion. Supported by the facilitators, they deliberated on the similarities and differences among the group reports. The same process was initiated again, with small groups deliberating on the trade-offs that they would be prepared to make in the implementation of their desired approach. The plenary reconvened to identify overarching commonalities. Participants then rated their level of agreement with the vision that they had collectively developed.

The CPRN's dialogues were not representative of the NWMO's much larger set of dialogues. Unlike other dialogues, those of the CPRN ran autonomously of the NWMO. For example, the CPRN hired a polling firm to constitute a statistically representative sample of the Canadian adult population to participate in its dialogues. In addition, it had Seaborn review its informational materials and workbook. Moreover, it documented the views of its participants in post-dialogue questionnaires. The CPRN's dialogues were by far the most systematically organized according to principles of deliberative democracy. Other dialogues, in which the NWMO played an active role in determining who would participate, presenting informational materials, and interpreting dialogue results, ran counter to the deliberative democratic aims of an informed and free exchange of public reasons among equals.[13]

A number of the NWMO's dialogues yielded significant measures of agreement. There were measures of agreement among members of the public sampled in the CPRN's dialogues (Watling et al. 2004). These participants agreed that safety for both existing and future generations is paramount and that it is the responsibility of existing generations to ensure that safety is attained. Participants also agreed on the merits of an adaptable and flexible management approach incorporating new knowledge and technology. They also agreed on the need for accountability and transparency. They argued that there "must be real engagement of experts, citizens, communities, and other stakeholders before any decision is made" (x). In the post-dialogue questionnaire, 77 percent of participants agreed with a waste management scenario incorporating these concerns (xi). Measures of agreement also emerged in the national and regional stakeholder dialogues. Participants agreed that used nuclear fuel is a significant risk to human health and the environment and that it needs to be safely managed for a long period of time (DPRA Canada 2004b, 7). They also agreed on the importance of a "neutral third party to oversee the development and implementation of the management approach" (24; see also Hardy, Stevenson, and Associates 2005a, 23). Participants generally agreed, moreover, that the waste management option must be socially acceptable and that there should be opportunities for any interested Canadian to participate in the process of selecting and implementing this option (DPRA Canada 2004b, 35; Stratos 2005, 5-6). These dialogues nonetheless highlight long-standing divisions among stakeholders about the place of nuclear within Canada's energy policy (DPRA Canada 2004b; Hardy, Stevenson, and Associates 2005a; NWMO 2005c; Stratos 2005; Watling et al. 2004). On this issue, participants could not agree and expressed strongly opposing views.

Based on its conclusions of the consultation process, the NWMO recommended to the federal government APM. This approach incorporates a step-by-step decision-making process, ongoing monitoring systems, and retrieval means for the irradiated fuel bundles. From the perspective of the NWMO, this approach is an amalgam of on-site storage, centralized storage, and deep geological disposal implemented over a protracted time frame. The approach incorporates the "unique strengths" of each of the three studied options. According to the NWMO, it "builds on the best features of the three approaches outlined in the Nuclear Fuel Waste Act and implements them in a staged or phased manner over time" (2005c, 23). Observers expressed support for the NWMO's recommendation for APM, claiming that the organization captured and addressed primary concerns articulated by participants (Berger 2005; Brook 2005; Hardy 2005; Member, NWMO Advisory Council 2005; Shoesmith 2005; Simpson 2005). According to Stratos, a "large majority" of participants in its stakeholder dialogues expressed "comfort with the recommendation" (2005, 4). Stratos claims that participants were "nearly universal" in their support for continuous monitoring over extended periods of time (5). Participants also supported sustained citizen engagement, public education, and community decision making (5). They also offered support for the NWMO's recommendation that a deep geological repository be "sited only in a willing host community" (6).

From another, more critical perspective, Adaptive Phased Management (APM) is problematic. According to Dave Martin of Greenpeace and Brennain Lloyd of Nuclear Waste Watch, this approach is "the worst of all worlds" (cited in Salaff 2005; see also Campbell 2005). They write that it "combines all the serious problems of at-reactor site storage of the waste, 'centralized' storage and deep rock disposal of the waste" (Campbell 2005). Perhaps even more problematic, APM seems essentially to be deep geological disposal over a realistic time frame. Arguably, any plans for such disposal would be phased over a comparative time frame. The NWMO's proposal seems to be a re-articulation of the industry's historical policy preference. It does not represent a substantive alternative to the industry's historical preference for deep geological disposal.

The Canadian nuclear energy industry has been advocating deep geological disposal since the late 1980s (CEAA 1998). During the 1990s, AECL and Ontario Hydro were the proponents in an environmental assessment review of conceptual plans for deep geological disposal in the Canadian Shield. The review process consolidated the opposing positions of two coalitions of actors. It polarized a coalition of industry and government officials against a coalition of environmental, religious, and Aboriginal organizations (Johnson 2007). The former argued that deep geological disposal is safe and should be publicly acceptable. The latter argued that it is not safe and not publicly acceptable. Critics argue that the NWMO's APM approach does not differ substantively from AECL/Ontario Hydro's concept of deep geological disposal of the 1990s (Harley 2004; Stanley, Kuhn, and Murphy 2004a, 2004b, 2005; United Church of Canada 2005a). They argue that the NWMO's APM approach is a blueprint for the actualization of the concept of deep geological disposal. It represents a plan for the siting and construction of a deep geological repository over a realistic period of time. Both finding a site for the repository and constructing it will be phased processes and will require decades of time. Nuclear scientists and engineers in the NWMO's dialogue on the technical aspects of nuclear waste management recognized the inevitability of on-site storage for the next fifty years even if the government opts for centralized permanent disposal (Shoesmith and Shemilt 2003, 4). They also noted that storage and disposal are not necessarily distinct but stages of an approach to permanent disposal allowing for flexibility and retrievability (5).

Conclusion

This analysis reveals questions about the extent to which the NWMO sought to realize principles of deliberative democratic citizen engagement in its national consultation process. The organization's process included a wide range of actors but not their perspectives. The process was characterized by a degree of procedural equality, but this equality did not extend to the equal consideration of all perspectives. It provided no opportunity for consideration of reduction at source. The NWMO's information sources were incomplete. According to Stanley, Kuhn, and Murphy (2005, 2), the NWMO was unjustifiably selective in its incorporation of views articulated in the consultations: "Without clear justification by the NWMO on how they sifted through the vast amount of material they collected ... and by what criteria they filtered that material, the role of the public in the process may be questioned, particularly by those who did not fully endorse the NWMO's conclusions."

The NWMO essentially prescribed the parameters of the dialogues, invited dialogue participants, and interpreted dialogue outcomes with a carefully circumscribed degree of transparency and accountability. As such, it reinforced the dominance of the industry coalition over the more critical coalition of environmental, religious, and Aboriginal organizations. Using its position of dominance in the consultation process, it tended to privilege the preference of the industry for deep geological disposal in terms of APM. In this way, the NWMO's process served to grant a questionable degree of legitimacy to its recommendation.

From an ethical perspective, deliberative democracy may be the most compelling ideal for decision making in areas of policy associated with risk, uncertainty, and conflict. Such areas are often contentious because health, societal, and environmental risks are high but characterized by great uncertainty. Deliberative democratic processes can render more comprehensive risk assessment, facilitate a reduction of uncertainty, and promote the reasoned discussion of implications by including a range of actors, as well as their perspectives, who could be affected by decisions in the policy area. Deliberative democratic practices can result in the ethical justification of decisions, which is especially important for policy that is risky and shrouded by uncertainty. In practice, the implications of deliberative democracy are far reaching, especially in the context of pre-existing power structures. This case reveals that there is much more to the deliberative democratic principles of inclusion, equality, and public reasoning than a prima facie implementation of them entails. Realizing the full extent of these principles requires the serious commitment of policy actors to the deliberative democratic ideal. In particular, it requires the serious commitment of members of dominant coalitions who tend to wield a disproportionate amount of power in given policy areas. Without such a commitment, the practice of deliberative democracy can take on a darker shade.

Notes

1 According to Seyla Benhabib, "legitimacy in complex democratic societies must be thought to result from the free and unconstrained public deliberation of *all* about matters of common concern" (1996, 68; emphasis added). For a discussion on the importance of wide inclusion in deliberative democracy, see Gutmann and Thompson (1999, 243-79; 2000, 144-64); Young (1999, 151-58).

2 For a discussion of the basic conditions of equality and freedom in deliberative democracy, see Chambers (1996); Cohen (1997a, 67-91; 1997b, 407-37).

3 For a discussion of the importance of epistemic equality, see Valadez (2001).

4 For a discussion of public reason in deliberative democracy, see Freeman (2000). For a discussion of reciprocity in deliberative democracy, see Gutmann and Thompson (2000, Chapter 2). See Valadez (2001, Chapter 2) on public reasoning in a multicultural context.

5 See Gutmann and Thompson (2000) on provisionally justified agreements. For a discussion of "incompletely theorized agreements," see Sunstein (1997). See Valadez (2001, Chapter 2) for a discussion of moral compromises.

6 Canada's nuclear reactors have generated about 2 million irradiated fuel bundles. If each of the reactors has an average operating life of over forty years, they will generate approximately 3.7 million used fuel bundles (NWMO 2005c, 15).

7 The NWMO played a direct role in inviting participants in all of the stakeholder dialogues: DPRA Canada (2004b); Hardy, Stevenson, and Associates (2005a); Stratos (2005).

8 The NWMO played a direct role in appointing members of or recruiting participants for the following roundtables, workshops, or dialogues: Carole Burnham Consulting and Robert J. Readhead Limited (Burnham and Readhead 2004); Coleman, Bright Associates, and Patterson Consulting (2003); Global Business Network (2003); Hardy, Stevenson, and Associates (2005b); NWMO (2004c); and Stratos (2005).

9 Numerous commentators confirm these findings. See, for example, Berger (2005); Brook (2005); Hardy (2005); Member, NWMO Advisory Council (2005); Mutton (2005); Rubin (2001); Shoesmith (2005); and Simpson(2005). See also DPRA Canada (2004b); Hardy, Stevenson, and Associates (2005a); and Stratos (2005). However, some observers held more critical perspectives. For example, a nuclear industry spokesperson thought that, because many of the dialogues were by invitation, certain individuals and organizations were excluded from participating (2005). See also Kamps (2005); and Stanley, Kuhn, and Murphy (2004a, 2004b, 2005).

10 See for example, AFN (2005d); DPRA Canada (2004b); Jackson (2005); Janes (2005); Kamps (2005); NWAC (2005); Nuclear Waste Watch (2004); Rubin (2001); Trudeau Foundation and Sierra Club of Canada (2005); and United Church of Canada (2004, 2005a, 2005c). See also Hardy, Stevenson, and Associates (2005a); Stanley, Kuhn, and Murphy (2005); Stratos (2005); and Watling et al. (2004).

11 See also Citizens for Renewable Energy (2005); Janes (2005); and Trudeau Foundation and Sierra Club of Canada (2005).

12 The NWMO held a workshop on the hazard of used nuclear fuel in February 2005. Its results were not posted until April 2005. Its objectives included assisting the NWMO in describing the types of hazard that need to be managed, identifying implications of this description for the NWMO's recommendation, and contributing to a balanced and scientifically sound portrait of the hazard (see Stratos 2005).

13 The NWMO made informational presentations to the following dialogues: Barnaby (2003); Burnham and Readhead (2004); DPRA Canada (2004a, 2004c); GBN (2003); Hardy, Stevenson, and Associates (2005a, 2005b); Sigurdson and Stuart (2003); and Stratos (2005).

7
Representing the Knowledges of Aboriginal Peoples: The "Management" of Diversity in Canada's Nuclear Fuel Waste
Anna Stanley

The recent history of nuclear fuel waste management has seen a dramatic increase in the political significance of Aboriginal peoples.[1] During the Seaborn hearings, testimony of Aboriginal peoples (mainly First Nations and Métis representatives) made public and explicit the ways in which many of their communities were and are implicated in the landscape of the nuclear fuel chain. Such testimonies narrated radically different regional and historical experiences of the effects of the nuclear fuel chain than did the narratives of proponents of the Atomic Energy of Canada Limited (AECL) concept of deep geological disposal and made apparent different judgments on the effects of nuclear waste and its management. The Nuclear Fuel Waste Act now requires that the Nuclear Waste Management Organization (NWMO) "consult" Aboriginal peoples on each of the management approaches under consideration. One manifestation of this requirement has been the NWMO's incorporation of the *knowledges* of Aboriginal peoples into its study.

The inscription of (previously invisible) Aboriginal peoples' geographies of the nuclear fuel chain onto the official politics of nuclear fuel waste management posed (at the time of the Seaborn hearings), and continues to pose, an important problem for the nuclear industry and NWMO. Significantly, it marks the introduction of the problem of "diversity" into the management of nuclear fuel waste. Nandita Sharma (2006) defines diversity as the "tangible existence of heterogeneity and mutual reciprocity," "born of lived *experience*" and "shared *practice*" (26, 150, 156; emphasis added). The NWMO, and indeed the coalition of nuclear actors and interests loosely termed "the nuclear industry," including owners and producers of nuclear waste such as AECL and the nuclear utilities, are now faced with a diversity of experiences, claims, and judgments – particularly those of Aboriginal peoples – many of which contradict and challenge their own claims. They must now create new policy narratives that somehow address the troubling, alternative, and unofficial knowledges and experiences of Aboriginal peoples of the fuel chain. Not only did Aboriginal peoples' judgments and experiences present a challenge to the narratives of the nuclear industry about the history and future safety of radioactive materials, but they also constituted (public) evidence of alternative knowledge and alternative (and negative) experience of the effects of radioactivity.

This chapter interrogates the appearance of the knowledges of Aboriginal peoples in the NWMO's work. My purpose is twofold: I wish to examine and comment on the ways in which the NWMO attempts to negotiate diversity (in the form of Aboriginal knowledges and experiences) and to demonstrate how the epistemic claims and contents of the knowledges of Aboriginal peoples relate to the management of nuclear fuel waste.[2] I suggest that the NWMO's study is in part an attempt to "manage" this diversity, contain the challenge that it poses, and fulfill requirements to consult Aboriginal peoples. The NWMO thus produces both difference and official diversity. Sharma has usefully defined difference as the "organized inequalities between human beings and between humans and the rest of the planet" as a result of "practices and beliefs founded upon hierarchies of differential value and worth" (2006, 26; see also Young 1990, Chapter 1). "Official diversity" Sharma describes as the banal reification of identity (as a natural cultural essence) in place of engagement with experience and lived practice (2006, Chapter 1). Both difference and official diversity, in the NWMO's work, cause homogeneity and the normalization of the (particular) claims, experiences, and perspectives of the nuclear industry rather than the actual inclusion of diversity (see Sharma 2006). I argue that the knowledges of Aboriginal peoples (and Aboriginal peoples themselves) are given a distinct place in the NWMO's process. They are named, circulated, considered, and therefore appear as markers of difference. However, the ways in which this body of knowledge actually appears serve to reinforce the positions and claims of the industry, by transcending Aboriginal knowledges (rhetorical incorporation), by obscuring them outright (absence), or by recontextualizing them (presence).

This chapter begins with a brief description of the ways in which the knowledges of Aboriginal peoples appear in the NWMO's work through *rhetorical incorporation, presence,* and *absence.* Focusing particularly on the ways in which Aboriginal knowledges are made a concrete presence in the NWMO's work, I demonstrate how, in order to apply and use concepts from Aboriginal knowledges in its study, the NWMO significantly transforms them. This results in applications of so-called traditional teachings that are significantly at odds with their intended meanings and that lead to (technical, social, and epistemic) directions in nuclear fuel waste management policy and planning that are opposed to the desires of Aboriginal peoples with respect to things nuclear. The NWMO's applications *differentiate* the knowledges of Aboriginal peoples from other (official) claims as something remarkable and work to obscure the diverse experiences of Aboriginal peoples of the nuclear fuel chain and thus to ensure that the effects of the fuel chain continue to be unloaded onto Aboriginal peoples' (primarily First Nations') lands and livelihoods. The NWMO's management of diversity ironically has sameness as its organizing principle and goal and serves to normalize the particular (and partial) claims of the nuclear industry.

Aboriginal Knowledge as it Appears in the NWMO's Work

The knowledges of Aboriginal peoples appear in the NWMO's work in specific ways. Two components make up the NWMO's study: the development of what it terms an "Analytical Framework" and the subsequent use of this framework (by a team of experts appointed by the NWMO and by two separate consulting firms) to evaluate the relative merits of the various management approaches and techniques under consideration. The NWMO's conclusions and final recommendation are based on the results of these evaluations. The response of various publics, including Aboriginal peoples, to the framework, overall process, and evaluations was solicited on various occasions throughout the process, particularly following the publication of the study results (e.g., NWMO 2003, 2004e, 2005a, 2005b). The development of the framework (including descriptions of it and of the process of developing it) was the most significant site of incorporation for the knowledges of Aboriginal peoples.

This framework is ostensibly representative of national citizen values, which are meant to encompass those of Aboriginal peoples. The framework was developed through a number of initiatives targeting Aboriginal peoples, including dialogues with Aboriginal peoples (wherein the five national Aboriginal organizations were funded by the NWMO to engage in dialogues with their constituent nations and governments about the NWMO's work), a two-day workshop on Aboriginal traditional knowledge (e.g., Barnaby 2003) held in September 2004, and an elders' forum established at the end of the process (25-27 August 2005). The different considerations that converge to constitute the framework and simultaneously represent citizen values are shown in Figures 7.1 and 7.2 (see NWMO 2005b, 47; 2004a, 67; and 2004e, 54). The "10 Questions" on the left-hand side of Figure 7.2 are a list of issues and concerns meant to encompass the questions that "Canadians" (including Aboriginal peoples) want answered in the management of nuclear fuel waste. The ethical and social framework, prominent in Figure 7.1, includes a set of seven citizen values and five Aboriginal values as well as four ethical principles that, again, are ostensibly representative of both Aboriginal and non-Aboriginal constituents. Surprisingly, the five Aboriginal values (honour, respect, conservation, transparency, and accountability) and seven citizen values (safety from harm, responsibility, adaptability, stewardship, accountability and transparency, knowledge, and inclusion) are remarkably compatible.

10 QUESTIONS

Institutions & Governance
Does the management approach have a foundation of rules, incentives, programs and capacities that ensure all operational consequences will be addressed for many years to come?

Engagement and Participation in Decision-making
Does the management approach provide for deliberate and full public engagement through different phases of the implementation?

Aboriginal Values
Have Aboriginal perspectives and insights informed the direction, and influenced the development of the management approach?

Ethical Considerations
Is the process for selecting, assessing and implementing the management approach one that is fair and equitable to our generation and future generations?

Synthesis and Continuous Learning
When considered together, do the different components of the assessment suggest that the management approach will contribute to an overall improvement in human and ecosystem well-being over the long term? Is there provision for continuous learning?

Human Health, Safety, and Well-being
Does the management approach ensure that people's health, safety and well-being are maintained (or improved) now and over the long term?

Security
Does this method of dealing with used nuclear fuel adequately contribute to human security? Will the management approach result in reduced access to nuclear materials by terrorists or other unauthorized agents?

Environmental Integrity
Does the management approach ensure the long-term integrity of the environment?

Economic Viability
Is the economic viability of the management approach assured and will the economy of the community (and future communities) be maintained or improved as a result?

Technical Adequacy
Is the technical adequacy of the management approach assured and are design, construction and implementation of the method(s) used in the management approach based on the best available technical and scientific insight? By method, we mean the technical method of storage or disposal of the used fuel.

ETHICAL AND SOCIAL FRAMEWORK
• Citizen and Aboriginal values and concerns
• Ethical principles
• Future Secenario
• Societal Context

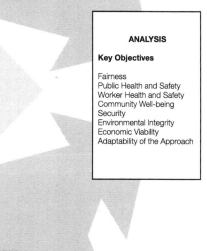

ANALYSIS

Key Objectives

Fairness
Public Health and Safety
Worker Health and Safety
Community Well-being
Security
Environmental Integrity
Economic Viability
Adaptability of the Approach

TECHNICAL INFORMATION
• Background papers
• Engineering design & cost estimates

Figure 7.1 The Components of the Analytical Framework. *Source:* NWMO 2005b, 45. *Reproduced with permission of the Nuclear Waste Management Organization*

These and other considerations were combined by the NWMO with the help of its Assessment Team to create the eight "key objectives" of the analytical framework, depicted on the right-hand side of Figure 7.2.

The knowledges of Aboriginal peoples, as a result of how they are incorporated by the NWMO into the analytical framework, appear in the work of the NWMO in three ways. The first can be described as a generalized rhetorical incorporation in the form of Aboriginal values. Here Aboriginal experiences and claims are represented as distinct and important values yet subsequently (and at the same time) rolled up as "values" consistent with those of (all) Canadians. This incorporation is rhetorical because it is persuasive and insincere (see Macey 2000, 329-30). It uses the presence of *potentially* different Aboriginal values (differentiated from other values always according to the NWMO's terms) to suggest diversity while paradoxically

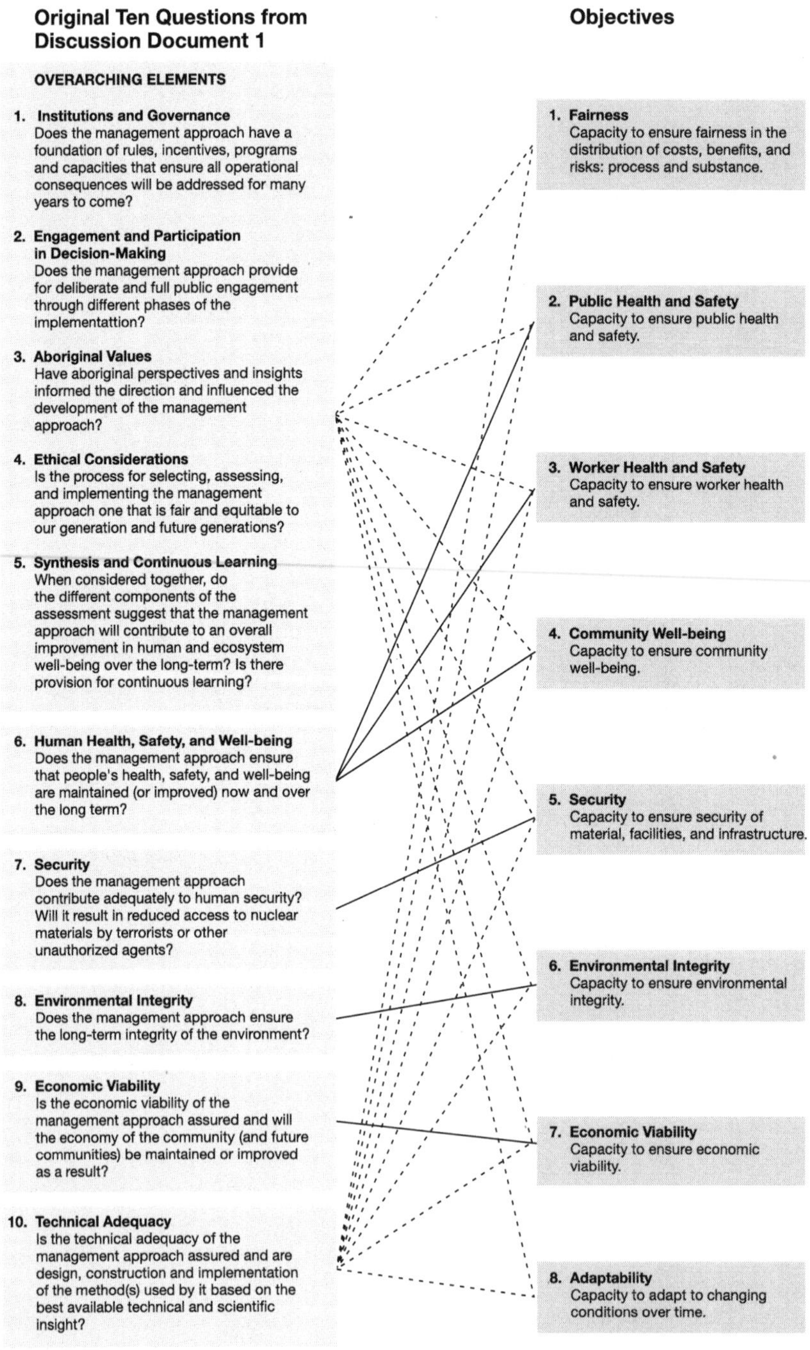

Original Ten Questions from Discussion Document 1

Objectives

OVERARCHING ELEMENTS

1. **Institutions and Governance**
    Does the management approach have a foundation of rules, incentives, programs and capacities that ensure all operational consequences will be addressed for many years to come?

2. **Engagement and Participation in Decision-Making**
    Does the management approach provide for deliberate and full public engagement through different phases of the implementattion?

3. **Aboriginal Values**
    Have aboriginal perspectives and insights informed the direction and influenced the development of the management approach?

4. **Ethical Considerations**
    Is the process for selecting, assessing, and implementing the management approach one that is fair and equitable to our generation and future generations?

5. **Synthesis and Continuous Learning**
    When considered together, do the different components of the assessment suggest that the management approach will contribute to an overall improvement in human and ecosystem well-being over the long-term? Is there provision for continuous learning?

6. **Human Health, Safety, and Well-being**
    Does the management approach ensure that people's health, safety, and well-being are maintained (or improved) now and over the long term?

7. **Security**
    Does the management approach contribute adequately to human security? Will it result in reduced access to nuclear materials by terrorists or other unauthorized agents?

8. **Environmental Integrity**
    Does the management approach ensure the long-term integrity of the environment?

9. **Economic Viability**
    Is the economic viability of the management approach assured and will the economy of the community (and future communities) be maintained or improved as a result?

10. **Technical Adequacy**
    Is the technical adequacy of the management approach assured and are design, construction and implementation of the method(s) used by it based on the best available technical and scientific insight?

1. **Fairness**
    Capacity to ensure fairness in the distribution of costs, benefits, and risks: process and substance.

2. **Public Health and Safety**
    Capacity to ensure public health and safety.

3. **Worker Health and Safety**
    Capacity to ensure worker health and safety.

4. **Community Well-being**
    Capacity to ensure community well-being.

5. **Security**
    Capacity to ensure security of material, facilities, and infrastructure.

6. **Environmental Integrity**
    Capacity to ensure environmental integrity.

7. **Economic Viability**
    Capacity to ensure economic viability.

8. **Adaptability**
    Capacity to adapt to changing conditions over time.

Figure 7.2 The Relationship between "Aboriginal Values" and the Framework.
Source: NWMO 2004e, 54. *Reproduced with permission of the Nuclear Waste Management Organization*

reinforcing the homogeneity of all values. The second can be described as presence: the knowledges of Aboriginal peoples are explicitly "written in" to the framework through a series of de- and recontextualizations of particular epistemological and moral teachings. Here the seven generations teaching, an epistemological structure of some (e.g., Anishnaabe) North American Aboriginal peoples' claims, serves as a proxy for all Aboriginal knowledge, but it is cast and recast in a manner compatible with the nuclear industry's and not, for example, First Nations', assumptions about and goals for nuclear fuel waste. Finally, this body of knowledge appears as an absence; Aboriginal knowledges and experiences are used, insinuated, and defined as a result of the ways in which their actual experiences are "written out" of the framework and hence of the official narratives framing nuclear fuel waste management.

Rhetorical Incorporation
In the transition from the preliminary ten questions to the eight objectives that ultimately constitute the analytical framework, the NWMO removed initial criteria that directly required it to reflect Aboriginal experiences, judgments, requirements, or concerns. Such removals eliminated specific consideration of Aboriginal values in the evaluation of a management approach (see Figures 7.1 and 7.2). The justification provided is that Aboriginal values, as seen through the NWMO's research, are the same as "Canadian" values. The analytical framework therefore claims that on a fundamental level Aboriginal and non-Aboriginal peoples hold the same values and articulate the same concerns with respect to nuclear fuel waste management: "Some aspects of the original Ten Questions are *generic* and apply across the assessment methodology ... This applies to the issue of Aboriginal Values (Question 3)" (NWMO 2004a, 68; emphasis added).[3] The generic character of Aboriginal values, as imagined by the NWMO and applied in its framework, is illustrated in Figure 7.2, showing the relationships between the original questions and the eight objectives of the framework.

The pattern of emphasizing yet ultimately dropping Aboriginal specificity, illustrated by the example above, is a significant trend in the NWMO's work. The particularity and special significance of Aboriginal "values," "perspectives," and "concerns" (as referred to by the NWMO) are consistently stressed and reported in exercises and publications describing the creation of the framework. The final report, for example, claims that "aboriginal peoples are an important community of interest for our study ... A number of contributions were also offered that reflect a special perspective that derives from the particular history, experience, and concerns of Canada's aboriginal community" (NWMO 2005c, 80-81). In the end, however, the NWMO stresses the uniformity of all contributions, emphasizing the unity between Aboriginal and all other (non-Aboriginal) values, perspectives, and concerns:

"Many of the observations and insights offered during the various elements of the aboriginal dialogues are similar to those gathered during our broader public dialogue. In particular: the highest priority concern is expressed for safety and security for people and the environment" (80). Far from representing an actual agreement or consensus, this integration of Aboriginal peoples' knowledges is rhetorical. It connotes the semblance of diversity while actually building consensus around a *particular* (Canadian[4]) rather than common or *universal* position. Here the difference created between Aboriginal and non-Aboriginal peoples along the lines of, for example, spiritual or cultural beliefs is about universalizing a particular, parochial set of interests: those of the nuclear industry.

This is a liberal social ontology and discourse about justice and difference. The NWMO's rationale for considering these diverse perspectives or experiences is to *transcend* the differences between them and to develop common ground, consensus, and "shared meaning" (e.g., Canadian Policy Research Networks [CPRN] 2004, 5-6). The knowledges of Aboriginal peoples (invariably called Aboriginal knowledge, Aboriginal traditional knowledge, and the perspectives of Aboriginal peoples by the NWMO) are primarily represented in the NWMO's work as one of a number of equally important and equally distinct citizen perspectives that must be considered to manage nuclear fuel waste. This treatment of diversity understands differences as a matter of (innate, natural) identity that *can* and *should be* transcended and therefore as something that it is *possible* to transcend (not in fact diversity). This understanding preserves the status quo (in this case the particular and partial claims and goals of the nuclear industry) and belies an important problem. Obscured are the ways in which the judgments and the ways of making judgments of Aboriginal peoples (in all their diversity) are rooted both in distinct ways of relating to and being in the world so as to produce claims about it and in distinct experiences of the nuclear fuel chain and its effects. Importantly, such an understanding obscures the ways in which the latter differ for structural and systemic reasons from those of non-Aboriginal peoples.

In this particular liberal discourse, difference is transcended and common ground is imagined and developed around various homogenizing tropes such as national identity, Canadianness, and citizenship (Stanley 2008). Such tropes provide convincing stories about a national "we" who experience equally the beneficial and positive effects of nuclear production. As I have argued elsewhere (2008), these are particularly powerful tropes around which this common ground is cultivated. It therefore makes sense that Aboriginal peoples' perspectives and experiences (once considered by the NWMO) appear totally consistent with, and indistinguishable from, "citizen values."

Absence

Despite the familiarity of many Aboriginal peoples, particularly First Nations peoples, with most parts of the fuel chain, its effects, and the geographical relationships that constitute it (see, e.g., Reckmans, Lewis, and Dwyer 2003), their knowledges and experiences do not form part of the NWMO's analysis. For instance, First Nations groups such as the Serpent River First Nation (SRFN), who have lived with the effects of nuclear industries, make informed and experience-based comments about the effects of long-term, low-level exposure to radiation from practices such as uranium tailings disposal or living near and working in uranium mines and mills. People of the SRFN comment, for example, on the effects of living in (hunting, fishing, swimming, eating, cultivating, and otherwise utilizing) an environment that they claim is contaminated by radioactive chemicals from past and present nuclear industry activities. Their comments pertain to the behaviour of radioactivity in ecosystems and its effects in and on food chains, ecologies, the watershed, and the geosphere of their lands and territories. They report instances of contaminated animals, illnesses, elevated cancer levels, and changes to water, plants, and medicines (Stanley 2004, 2006). Furthermore, they make claims about the effects of nuclear industries on their societies and their (in)ability to practise their Aboriginal and treaty rights (e.g., Federal Environmental Assessment Review Office [FEARO] 1993, 1996; Reckmans, Lewis, and Dwyer 2003). This knowledge is absent from the NWMO's evaluation, framework, and definition of Aboriginal peoples' contributions to nuclear waste management not simply because it is (deliberately or otherwise) omitted or overlooked. These knowledges are an "absent presence" in the NWMO's work.

I call this appearance "absence" because Aboriginal knowledges of the fuel chain and its geography are literally "written out" as a result of the ways in which the role of Aboriginal knowledge and its content are presented in relation to nuclear fuel waste management in the NWMO's account. Here the relationships between Aboriginal peoples and the fuel chain, and between Aboriginal peoples and knowledge production, are structured by the NWMO in ways that preclude experience of the fuel chain. Aboriginal peoples' claims about the effects of nuclear industries, including the long-term, low-level effects of radioactivity on human bodies and the environment, are de facto excluded because of how Aboriginal peoples and their knowledges are unconnected to the fuel chain. Aboriginal peoples (as a category) are never in the NWMO's work associated with specific geographical locations or places and never connected to sites related to the nuclear fuel chain. Moreover, the NWMO never mentions the nuclear legacies in Aboriginal communities or on Aboriginal lands (Stanley 2008). A number of narratives and discursive strategies normalize the disconnection of Aboriginal peoples from the fuel chain, and here I briefly describe two.

Strategy 1: The Essentialization of Aboriginal Knowledges

Aboriginal knowledge is described in the NWMO's work as both an "object," or data to be collected and used by the NWMO to help it come to conclusions, and a series of abstract teachings (e.g., the seven generations teaching) about procedural matters. Both represent Aboriginal knowledges about the nuclear fuel chain as something constructed not through experiences and familiarity with radioactivity and nuclear materials but through aboriginality and natural cultural inclinations. Both create a relationship between Aboriginal peoples and nuclear fuel waste management that has nothing to do with waste but with identity (as intrinsic values and cultural beliefs).

The NWMO objectifies Aboriginal knowledge by framing it as something that it can "utilize" (e.g., in future phases such as concept implementation or siting). Describing the knowledge of Aboriginal peoples in relation to its study, the NWMO frequently refers to it as including "special knowledge related to the land" (2005c, 83) and portrays it as data and information about a specific landscape. This objectification of Aboriginal knowledge presents it as data suitable for assisting the NWMO in characterizing a site location, negotiating with a particular community, or developing impact mitigation strategies and thus as subordinate to scientific or Western rationality (Stanley 2006). As the NWMO states, "we have begun the process of learning how to integrate the insights and knowledge of aboriginal peoples into our work. There is substantive knowledge about the land and ecology in any given location, stemming from long contact with the land. But Aboriginal Traditional Knowledge is also about ways of developing and maintaining effective and respectful relationships, between young and old, within a community, between communities" (NWMO 2005c, 19). This is a narrative that constitutes Aboriginal knowledges as distinctly local: as data about a specific place that is distinctly bounded, as knowledge that is parochial (bound up with local interest), and as knowledge that is epistemologically bounded – that is, does not travel to other places and requires universal frameworks to rationalize and make sense of it (Stanley 2006).

Aboriginal knowledges are also portrayed as abstract and procedural. These representations essentialize Aboriginal knowledge by constituting it as a natural attribute of aboriginality and as unrelated to shared practice and experience of the nuclear fuel chain. It is described most often as "process related insights" (e.g., NWMO 2005c, 83), desirable because they are *Aboriginal* insights. As a result, this knowledge is naturalized, represented as an innately cultural knowledge. For example, the NWMO states that it has "learned that aboriginal traditional knowledge includes both an understanding of nature and of all human relationships. It sees humans as part of the environment and spirituality a component of all relationships" (2005c, 82). Nowhere in this passage is Aboriginal knowledge connected epistemologically to experience of the nuclear fuel chain. Indeed, it is normally presented as largely irrelevant to nuclear fuel waste management except insofar as the NWMO design to apply it (with the help of Western science) to the management of nuclear fuel waste. In its final report, for example, the NWMO describes the application of Aboriginal knowledge: "Applying these principles [from Aboriginal knowledge] (for example, in any NWMO process) would involve the elders and wisest speaking first, praying for assistance to make good decisions, constantly growing and evolving new insights, involving the whole community" (2005c, 83). This description essentializes the knowledge by removing its basis in experience and separating it from epistemic contact with the nuclear fuel chain. It becomes an essential (cultural) characteristic of Aboriginal peoples that it is important to include in order to take advantage of the "special insights" of Aboriginal peoples. This representation of their knowledges obscures the negative experiences of the nuclear fuel chain on which claims about it are based. Furthermore, it disconnects Aboriginal peoples from exposure to, experience of, and judgments about the effects of nuclear industries and radioactivity.

Strategy 2: Aboriginal Peoples as an Abstract National Community

An important set of spatial representations and associations connected to the creation of an imagined national landscape of the nuclear fuel chain silences the knowledge of Aboriginal peoples (Stanley 2008). These representations subtly establish relationships between Aboriginal peoples and the nuclear fuel chain that write out all evidence of experiences (and therefore knowledges) of the fuel chain. Aboriginal peoples are constituted as an abstract and unified group unspecified in terms of place, with the exception of national space. As such, they are never mentioned in connection with any location (region, city, community) associated with the nuclear fuel chain, nor are nuclear activities on their lands ever given material expression in the landscape of the fuel chain. Aboriginal peoples are instead connected to "lands," "communities," and "traditional territories" as abstract concepts as well as to "Canada" and the space of the nation. Mention is made of Aboriginal peoples' relationships to the land, and of "Canada's Aboriginal community," "Canada's Aboriginal peoples," and "Aboriginal Canadians" (e.g., NWMO 2004e, 21-22; 2005b, 163; 2005c, 80-82), but never, for instance, to the Dene of Delené the Serpent River First Nation near Elliot Lake, or the Chippewas of Saugeen and Nawash on the Bruce peninsula – just some of the First Nations whose lands and livelihoods have been implicated in the geography of the nuclear fuel chain. As a result, Aboriginal peoples are constituted in the NWMO's work as important because they *are* Aboriginal, and their knowledge because it is reflective of aboriginality, not because they have experiences of nuclear activities or the capacity to make informed judgments about them.

Both of these strategies result in the acute essentialization of Aboriginal knowledge as an object subordinate to science, a (natural) attribute of aboriginality, and a phenomenon unconnected to experience. They work to constitute a relationship between Aboriginal knowledge (and Aboriginal peoples) and nuclear waste management that obscures and de facto writes out the history of diverse and regionally varied experiences of Aboriginal peoples of the nuclear fuel chain and the epistemological basis of knowledge claims and judgments about the effects of nuclear industries. Thus, the knowledges of Aboriginal peoples about the effects of radioactivity and chemical toxicity, and their claims about its health, environmental, social, and ecosystem effects, are made absent from the NWMO's study in favour of official and uncontroversial ones.

Presence

Beyond rhetorical incorporation and absence, which insinuate diversity and circulate Aboriginal difference in its work, the NWMO also applies and uses the knowledges of Aboriginal peoples concretely throughout its framework. This gives Aboriginal knowledge a distinct presence in nuclear fuel waste management and for this reason makes a convincing suggestion of diversity and inclusion in the NWMO's work. I call this third and perhaps most important way in which the knowledges of Aboriginal peoples appear in the NWMO's work in relation to nuclear waste "presence," where Aboriginal knowledge is directly handled by "writing it in" to nuclear fuel waste management. The seven generations teaching is the one instance of presence in all of its work. It is the only concrete application and use of "Aboriginal knowledge" in its study. Although there are many references to the special relationships that Aboriginal peoples have with their lands, Aboriginal spirituality, and so forth, in NWMO publications the only concept that is actually applied is the seven generations teaching. It therefore becomes a *proxy* for all "Aboriginal knowledge" through the NWMO's work:

> Traditional knowledge is more than a simple compilation of facts drawn from local, and often remote, environments. It is a complex and sophisticated system of knowledge drawing on centuries of wisdom and experience. It constantly grows and changes with new information ... Traditional knowledge systems assume that people are part of the land, not that they own the land, so they consider themselves as true guardians. Traditional ecological knowledge emphasizes the inter-relationships between components of the environment and avoids scientific reductionism. *The wisdom derived from this philosophy can be used when planning for the future. For example the "seven generation" teachings require decision makers to consider the impact of their choices on future generations.* This wisdom if applied, may contribute to ensuring that a management approach for used nuclear fuel in Canada is sustainable in the long term. (2003, 47; emphasis added)

That the NWMO conflates Aboriginal knowledge with the seven genera-tions teaching is itself problematic. Here, however, I want to focus on how, in order to utilize the concept in its analysis, the NWMO transforms it to make it consistent with NWMO assumptions and goals. In so doing, the NWMO applies it in ways that shift and alter its meaning. Through a series of decontextualizations and recontextualizations (of which I identify three sets), it transforms the meaning of the seven generations teaching away from the ways in which it is applied to things nuclear according to Aboriginal people's knowledge-making practices. Here I interpret the teaching mainly according to the knowledge practices of First Nations peoples, such as the Serpent River Anishnaabe and the many First Nations who participated in the AFN dialogues on nuclear fuel waste, established as part of the NWMO's Aboriginal dialogues.[5] These transformations make Aboriginal knowledges appear to be consistent with the knowledge claims of the nuclear industry. They also legitimize certain ways of handling, for example, uncertainty and managing nuclear waste that are (even according to the NWMO's own re-search) at odds with the preferences of Aboriginal peoples.

Recontextualization 1

The NWMO most consistently uses this teaching as a request and directive from Aboriginal peoples to take into account and consider the future effects (social, environmental, and so on) of nuclear waste management approaches, reporting that "they [Aboriginal peoples] have advised us to think of the impact of our actions seven generations hence" (2005b, 48). Indeed, em-phasizing its futuricity, the NWMO often does not limit itself to the specified time period of seven generations when applying the teaching: "The seven generations teaching and its inherent consideration of impacts *many genera-tions out,* has greatly influenced the NWMO study process" (2004e, 22; emphasis added). In this application, the teaching has been taken radically out of context. By contrasting this application with the ways in which the SRFN (an Anishnaabe nation), with extensive experiences of radioactivity, radiation, and the nuclear industry in its lands, uses it in relation to things nuclear, the NWMO's transformations of the concept become apparent.

At a public hearing in 1993 at the SRFN into a uranium mining company's proposed uranium tailings management plans for the Serpent River Water-shed, an elder of the SRFN employed the seven generations teaching to the management of radioactive uranium tailings (waste rock) in the following way: "I find this a very difficult task to do because within our Indian ways, we are taught that all our decisions should be made by thinking *seven genera-tions in advance,* that when I'm here to speak for my family, my thoughts should be projected *seven generations in advance*" (FEARO 1993, 99; emphasis added). He went on to say that "we talk about this waste that is going to be here for a very long period of time. We know that the period of time I believe they talk in terms of half lives, they talk in terms of tens of thousands of years. There is a lot of time before this stuff becomes neutralized, or this stuff becomes harmless again. Now this process that is going to take place *over several thousand generations* before this product is harmless again. Is this panel prepared, or are they even capable of making decisions *for that length of time?"* (105; emphasis added).

The contrast between seven generations and several thousand generations is significant. Through this contrast, the elder compares the conceptual limits that the SRFN places on knowledge making (seven generations) with those of the nuclear industry (several thousand generations or more). He therefore uses the teaching to question a uranium mining company's ability to make judgments about the safety and acceptability over time of the low-level nuclear wastes that will be disposed of for eternity in the Serpent River Watershed. In this case, both the company and the federal regulator claim that the tailings containment method will be safe because the risks that it poses to the ecosystem and humans will remain within regulatory levels for *eternity.* In using seven generations in this way, the elder is being ironic: he is challenging the industry's seemingly unlimited ability to produce knowledge and particularly its ability to produce knowledge so far outside the limits of several thousand generations. He is therefore questioning the quality and content of the industry's claims about the future. The message implicit in the elder's application of the teaching is that there are distinct limits to the future application of time-space experience and hence to the production of knowledge based on today's experience (see also AFN 2004a).

This juxtaposition indicates that the seven generations teaching, as applied by the NWMO, has been decontextualized from its intended meaning (as practised by the SRFN). The "limiting" condition associated with the teaching has been removed, therefore eliminating the epistemological statement about the boundaries of knowledge production and space-time limits of experience that forms part of the teaching and, importantly, eliminating criticism of the content and method (implicit in the teaching) of the nuclear industry's claims about the future. This leaves only the portion of the teaching that focuses on the future implications of a decision. The teaching is then recontextualized as a teaching that has an unlimited extension into the future and as an important directive instructing the NWMO that it should and can make knowledge, within an unlimited time-space horizon, about nuclear fuel waste.

Recontextualization 2

Speaking once again in relation to the management of radioactive uranium tailings, an elder from the SRFN makes the following statement in which the seven generations teaching is applied to the production of knowledge about radioactivity. In an earlier part of his testimony, he introduces the concept of seven generations. He then states,

> I would like to say a few words about the industry itself, the nuclear industry itself. As we all know, the industry starts from the actual mining process, and this is one thing we are dealing with today, a residue of the mining process. After the mines are put in place and the ore has been taken out of the ground, it then goes to a refining stage. This refining stage requires a lot of toxic, highly toxic chemicals to be induced into this pot-pourri or this mixture, in order to take out the element that you are looking for. Once this process has taken place, there is a lot of toxic waste by products which are being left around for us to deal with ... We then have to worry about the utilization of this particular mineral or this particular resource, in particular the nuclear fuels. Once this fuel has been used up, we now have to be concerned with the disposal of this waste. (FEARO 1996, 761)

Having described at length the nuclear fuel cycle and its relationship to the waste and management problem, the elder continues: "They still do not know how to look after the processes that they have created. I believe that the nuclear industry has failed in dealing with all of these elements, because they have failed the test of time. They have not been able to create an industry which would be safe for the people that would *span a period of seven (7) generations*" (761; emphasis added).

Here the elder is using the teaching to direct attention toward the production and life cycle of the waste. He is also directing attention toward the decision to generate nuclear power and mine uranium and in so doing creating a connection between management of these things and management of the waste itself. The message from this intervention and use of the teaching is that the waste cannot be managed in isolation of its production and that, in historical terms, the decisions to mine uranium and generate nuclear power need to be revisited and, in his opinion, addressed. He is using the teaching to connect safe waste management to the production of the waste, insinuating that we return to the original decision to create the waste or safety cannot be ensured.

When compared with the NWMO's application of the teaching in the first example, it is again clear that the teaching has been taken out of context by the NWMO. The teaching has been decontextualized by removing its reflexive and historical qualities that would shift the focus to include the entire fuel cycle when considering the management of its wastes. It has also been recontextualized as an ahistorical teaching that can be applied to the management of the waste as an end product only, in isolation of the social history of the waste, the production cycle, and the decision to produce the waste. This particular recontextualization has also been noted by the Assembly of First Nations, which, in its report to the NWMO under the heading "Inappropriate Characterization and Use of Aboriginal Traditional Knowledge," scathingly notes that "of particular concern was the inappropriate use of 'seven generations' teachings ... The working group felt that the inappropriate use of seven generations teachings was done simply to appear that the NWMO was incorporating ATK [Aboriginal traditional knowledge] into its work rather than looking critically at what those teachings truly say about the *production* and management of used nuclear fuel" (2004a, 7; emphasis added).

Recontextualization 3
The following passage taken directly from the NWMO's analytical framework represents the third way in which the NWMO transforms the meaning of the teaching. In its analytical work assessing the differences between the different management approaches, the NWMO analyzes the options within two distinct time frames: the present to 175 years, and 175 years to 10,000 years. These time frames are justified in part with reference to the seven generations teaching. The time frames are defined as follows:

> Period 1. From the present until 175 years from now: This period roughly corresponds to the "seven generations" used by Canadian aboriginal peoples as a target or goal for assessment or evaluation of benefits or consequences of current issues.
> Period 2. Beyond 175 years: Beyond seven generations and up to 10,000 years. Aboriginal perspectives and future scenarios work conducted by NWMO suggest that continuity from the present conditions and present situations *cannot be assumed socially, institutionally, or environmentally.* Although *geological characteristics can be predicted with some confidence,* the vagaries of physical environmental conditions and human induced or natural stresses *on the ecosystem* make any assessment of the *human ecological interactions* extremely speculative. (NWMO 2004a, 69; emphasis added)

Compared with both passages from the SRFN, this constitutes a third way of transforming the teaching from its intended use by First Nations in knowledge production. Although similar to the initial set of de- and recontextualizations, it represents a subtle and important difference, one crucial to justifying the NWMO's handling of uncertainty and choice of management approach.

The seven generations teaching is decontextualized in the passage above by once again removing the limiting condition, as was the case earlier, and by keeping only the forward-looking directive to plan for the future. This tactic is particularly evident in the definition of the first "time period." In this instance, however, the teaching is recontextualized by adding a different and weak version of the limiting factor, one that is not consistent with the epistemological boundaries put in place by the Serpent River First Nation and others. This weak factor places limits on our ability to know about the future not because knowledge is uncertain but because the things that we want to know about are *themselves uncertain*. For example, in the second part of the passage, the ecosystem, human institutions, and the future are all represented as *inherently* uncertain things. The teaching is then recontextualized by rather arbitrarily applying this weakened limiting factor disproportionately to social, institutional, and environmental continuity with the present and not to other characteristics of the future, such as geological, hydrological, or chemical continuity, or to human-engineered scientific objects, or certainly not to the ability to produce knowledge about them over time.

This series of recontextualizations (because they are embedded in the analytical framework) helps to justify particular policy outcomes and analytical results that are inconsistent with the stated positions of organizations representing Aboriginal peoples (e.g., AFN and the Native Women's Association of Canada [NWAC]) regarding nuclear fuel waste management. This is particularly the case with respect to the NWMO's assumptions about uncertainty.

The NWMO has recommended Adaptive Phased Management (APM), an option that specifies deep geological disposal as the final containment method implemented over a time frame of up to 300 years, with a possible intermediate step of centralized storage prior to moving to final disposal (2005c, 29). In the interim, waste will remain at reactor sites. Monitoring and retrievability are claimed as possible once the repository is sealed. Furthermore, the integrity of the system, as a passive containment system, relies on the repository being completely sealed and closed off to ensure safety, security, and isolation of the waste. This method is preferred by the NWMO on several grounds, justified in its analysis. First, it is a *passive* approach. Although it provides a 60- to 300-year delay before moving to an eventually sealed deep geological repository, the repository, once sealed, does not rely on human institutions, processes, or actions to ensure safety; rather, it relies on human-engineered scientific objects such as concrete walls and metal canisters and, more importantly, on physical environmental barriers (negative pressure at a certain depth and a saturated vault) to isolate waste in a safe manner (e.g., 2005b, 31). Second, it is the most *certain*; it is presumed by the NWMO that the final stage of deep geological disposal is the most sure way to isolate waste over time and to reduce risk, given the uncertainties surrounding nuclear fuel waste (e.g., 2005b, 7). Third, it is flexible in the face of all possible future energy scenarios and therefore does not presume energy policy or the future of nuclear energy (e.g., 2005b, 26). The NWMO's application of the seven generations teaching favours these features and is consistent with claims made by the owners and producers of nuclear wastes, the assessment team, and the NWMO in its final study that they can estimate with confidence the safety, security, and general performance of the repository over tens if not hundreds of thousands of years (e.g., 2005c, 45, 221). The teaching as reformulated is consistent with the nuclear industry's preference for passive over active methods of containment and with its unique focus on the management of the waste and its future effects rather than its past, present, and future effects. It also obscures the contributions of Aboriginal peoples, in terms of both epistemology and experience, to nuclear fuel waste management and the important challenges that they present.

The NWMO's decontextualization of seven generations justifies a particular negotiation of uncertainty and, importantly, the judgment that the uncertainty (especially long-term uncertainty) associated with nuclear fuel waste management can be successfully handled using (scientific) methods of modelling and prediction. Specifically, the recontextualized teaching normalizes the NWMO's dissociation of uncertainty from the *production* of knowledge and its subsequent association of uncertainty with the *things* that it wants to know (Stanley 2006). As a result, uncertainty is no longer a characteristic of knowledge production, as it is in First Nations' use of the teaching, but becomes an inherent property of things, such as the future, society, the environment, and so on (Stanley 2006). As reformulated by the NWMO, the problem faced by society is not knowledge production about the future but the future itself. Further, certain knowledge production techniques (read science) can handle (and in some cases transcend) uncertainty to make superior knowledge about the future. As such, the seven generations teaching is used as evidence that it is desirable (and perhaps possible) to transcend uncertainty and to support the claim that knowledge can and should be made, without time-space limits, about the future: "We are contemplating designing and licensing a system that would last for periods longer than recorded history ... What we can do is plan for the foreseeable future, act responsibly and confidently with the best science and technology in hand" (NWMO 2005b, 12-13). These types of assumptions build into the framework a comfort with knowledge claims that rely on complex structures of prediction, and risk and probability narratives about the future, and therefore support for methods that overwhelmingly rely on the ability of nuclear scientists (armed with today's empirical experience) to predict repository performance. Given the long time frames over which safety must be ensured, it is, according to the NWMO, legitimate (indeed preferable) that "detailed scientific studies, models and codes form the foundation of the assurances of performance provided to regulatory authorities and interested organizations and individuals" where experience and empirical evidence are not possible (2004e, 8).

Furthermore, the distinction between "societal" and "technical" uncertainty made by the recontextualized teaching normalizes the NWMO's assertion that societal aspects of nuclear waste management (including, interestingly, the biological environment) are more uncertain in the long term than technical and scientific aspects (including knowledge production itself and the physical environment)(e.g., 2005c, 220). Following from the assertion by the NWMO that "technical and scientific" realms (including some environmental characteristics such as hydrology or geochemistry and geophysics) are inherently more certain and their continuity from the present somehow ensured, and that knowledge making is not the problem, the NWMO justifies its recommendations based on an assumption that social institutions cannot be relied on, whereas the continuity of scientific and technical realms as well as scientific and technical knowledge can be (e.g., 2005c, 24, 220). This claim (that social institutions cannot be relied on, whereas the continuity of scientific and technical realms as well as scientific and technical knowledge can be) enables the NWMO to build into the framework assumptions that disproportionately associate uncertainty and the effects of time with societal and environmental characteristics. Privileging this association suggests that what is required is a passive management approach, which relies on the continuity of technical and scientific properties, such as geology, chemistry, and hydrology, and assumes the ability to produce credible knowledge far into the future about these properties, the behaviour of radioactive materials, and the safety and performance of an engineered containment system. Justifying its recommended approach (described above) with specific reference to the criteria in its framework, the NWMO

> believes that the type of responsible and prudent approach that Canadians have said is required dictates that we *not* rely on the existence of strong institutions and active management capacity over thousands and tens of thousands of years. On this basis the NWMO does not suggest either of the storage options as a preferred approach for the long term ... This approach [APM] clearly identifies the technology associated with a deep geologic repository as the appropriate end point. It does not rely upon human institutions and active management for its safe performance over the long term. (2005c, 30-31; emphasis added)

Faced with criticism that it is neglecting the source of the waste in conceptualizing a method of managing nuclear fuel waste, the NWMO has repeatedly responded that the production of waste, generation of electricity, and fuel chain are outside its mandate (e.g., 2005b, 41; 2005c, 20). The NWMO's reconceptualization of the seven generations teaching also normalizes the argument that waste can be managed in social, historical, and physical isolation from the fuel chain regardless of the future of nuclear production and with no bearing on the future of this industry. Certainly, the NWMO is not required by the act to address the production of nuclear fuel waste, energy policy, or the future of nuclear power. However, despite this mandate, not considering the generation of the waste as part of waste management constitutes, many have argued, a tacit agreement that the management of waste and the production of energy are and can be separate and that it is acceptable for them to be considered in isolation. The NWMO further maintains that a nuclear fuel waste approach and a management option have no bearing on the future of energy policy or a renaissance of nuclear production (e.g., 2005b, 26). Ignoring the historical and reflexive dimensions of seven generations assists its ability to maintain this fiction while focusing analysis and debate on the potential future impacts of the approaches rather than on the past decisions taken about nuclear fuel waste management and the policy and historical connections between waste production and management. This recontextualization puts in place assumptions in the framework about nuclear fuel waste that help the producers and owners of nuclear wastes (and other nuclear industry actors) to continue to avoid the reformulation of nuclear fuel waste in terms of the fuel cycle and allow the NWMO to apply "Aboriginal knowledge" within the comfort of what it interprets as its mandate.

As a result of this recontextualization, the NWMO uses the teaching throughout its evaluation and study to support a particular handling of uncertainty. This in turn leads the analysis to favour approaches to and techniques for the management of nuclear fuel waste that are, even according to the NWMO's own research, opposed by Aboriginal peoples. As I document below, the management options that Aboriginal peoples overwhelmingly prefer, according to both their own submissions and NWMO documents, are critical of guarantees of safety over the long term and show a distinct preference for active management approaches (especially over the long term) where waste is constantly monitored, retrievable, and never far from reach or sealed and for the phase out of nuclear power as part of the management option. These preferences are not consistent with the management option that the NWMO recommended at the end of its study, which, as I have documented above, relies for proof of its long-term safety on modelling and complex probability and risk assessment to predict behaviour, prefers burial deep underground to ensure long-term safety, and arguably supports monitoring and retrievability only until final decommissioning.

According to the NWMO, Aboriginal organizations participating in its process overwhelmingly opposed deep geological disposal options, favoured monitored, retrievable, near-surface storage options, expressed little faith in the ability of scientific and technical prediction over the long term, and insisted on the necessity of considering (if not phasing out) nuclear production, energy policy, and the production of the waste as part of the management strategy. The phasing out of nuclear power was particularly emphasized. Indeed, even those few organizations (e.g., Inuit Tapiriit Kanatami [ITK], the western Indian Treaty Alliance, and the Ontario Métis Association) that advanced cautious and qualified support for the NWMO's final recommendation did so based on the caveat that nuclear power generation cease and that alternative energy sources be actively sought (e.g., NWMO 2005c, 110). According to the NWMO's final report, "many aboriginal participants" favoured reducing nuclear energy and "argued that the waste management issue cannot be resolved without a broad discussion of energy policy" (80; see also 81, 101, 110). The NWMO also reports that Aboriginal peoples argued for surface or centralized storage near large population centres to ensure continued monitoring (88, 96; see also 81, 97, 99) and stressed flexibility and adaptability in a management strategy (97, 99). Such concerns led the NWMO to admit that "many aboriginal peoples suggested they did not accept the deep geologic storage" (101; see also 10, 81, 87).

A review of the submissions made by Aboriginal organizations (primarily First Nations organizations and governments) reveals these positions to be even more pronounced. NWAC, for example, stated in its submission to the NWMO that it did not support the NWMO's recommendation because "there is no assurance that the production of more and more nuclear waste will be curtailed in Canada any time soon" and that "there is no discussion of pursuing alternative, 'green power' options" as part of a management approach (2005, 5). Similarly, the AFN,[6] in all of its reports on regional consultations with constituent First Nations, states that it is necessary for any nuclear waste management strategy to consider production of the waste: "Participants felt strongly that considering the entire nuclear energy cycle was the only way to properly discuss the nuclear waste issue" (AFN 2004c, 4) and that "the first step in dealing with the management of nuclear fuel waste was to immediately stop its production" (2005e, 5; see also AFN 2004c, 2004d, 2005b, and 2005c). This position was also adopted as a resolution at the AFN Annual General Assembly in July 2005 on behalf of all First Nations (AFN 2005a).

For various reasons, regional forums of First Nations rejected methods involving deep geological disposal or burial (including APM). According to reports by the AFN, constituent First Nations thought that "the management options, particularly deep geologic disposal, are naïve in their whole hearted acceptance of science and technology and in understanding the environment in which we all live," including assumptions pertaining to risk methodologies (2004c, 5). In particular, they expressed concern with the time frame over which nuclear waste remains harmful and over which it would be necessary to ensure safety, stating that "it was not possible to ensure the waste would be contained for the duration that it was hazardous to future generations" (5). Many First Nations consulted by the AFN found options based on deep geological disposal (including APM) particularly "risky" and uncertain over the long term, specifically because of the type of prediction and knowledge on which they relied to ensure safety and containment (7, 9; see also AFN 2004b, 5). Specifically on this issue, First Nations consulted at a regional forum in northern Ontario made reference to the seven generations teaching to point out the uncertainty of deep geological disposal. As reported by the AFN (2004b, 6), "the participants described the indigenous concept of looking seven generations ahead in decision making processes. Decisions made today will be inherited by the next generation and they will in turn think of the next seven generations to come ... Participants were very concerned over the length of time nuclear fuel waste will remain hazardous to humans and the environment. They felt that it was not possible to guarantee that it could be contained for the entire time that it was hazardous to future generations. The inherent responsibility to protect lands within their territories was a driving concern."

Several observations can be drawn from the contrast between First Nations' and the NWMO's interpretations of the seven generations teaching described above. The teaching, when applied to the management of nuclear fuel waste or other nuclear policy issues, is used as an important critique of both the content and the claims of the knowledge production techniques preferred by the nuclear industry and supported by the NWMO in its study. This is a teaching that puts around knowledge production strict boundaries that limit the time-space reach of claims, especially claims about the future. This leads to a radically different treatment of uncertainty from that accomplished through the NWMO's analysis and written into its framework. Uncertainty, according to First Nations' knowledge practices, is a condition of knowledge produced about the future and therefore cannot be transcended by Western techniques such as universality or objectivity. Nor is uncertainty more pronounced for the production of knowledge about social versus physical things. These applications of the teaching favour methods that accommodate uncertainty (such as constant monitoring), insist on reliance on human interventions, emphasize the importance of human and social institutions, and keep the waste near the surface for constant surveillance and active safety. These applications are therefore not consistent with reliance on passive safety methods. Most importantly, First Nations' use of the teaching insists that the stakes are high enough, and the uncertainty great enough, that waste production must cease and that efforts should focus only on monitoring and managing what waste has been produced. By recontextualizing the seven generations teaching, the NWMO depoliticizes the claims made by Aboriginal peoples, dismisses the critiques that they contain, and allows the organization to use "Aboriginal knowledge" in its study in ways consistent with the goals and plans of waste owners and producers and other nuclear industry actors.

Concluding Remarks: Difference and Diversity

Since the Seaborn hearings, the nuclear industry has been faced with the problem of diversity – in the form of alternative claims and experiences of Aboriginal peoples concerning the nuclear fuel chain. The NWMO's representations of the knowledges of Aboriginal peoples are, I have argued, an attempt to manage diversity in favour of the nuclear industry and contain the challenge that they present. This conclusion is consistent with theoretical insights from Sharma (2006) and Young (1990) about social justice and difference. The three appearances reviewed above produce "official" diversity because they mark out, specify, and parade Aboriginal knowledge as an important and significant contributor to the study according to differences manufactured by the NWMO. Aboriginal diversity is represented as special perspectives and distinct cultural attributes (e.g., spirituality, tradition) resulting in the conflation of diversity and innate characteristics rather than experience and practice. The NWMO does not recognize the "real" diversity, which in the case of Aboriginal peoples results from the disproportionate ways in which they have been implicated in the fuel chain, the different experiences (and knowledges) that they therefore have of it, and the distinct practices and histories of the relationship to people and the environment through which they experience and make judgments about it. Diversity thus becomes a problem for the NWMO of accommodating Aboriginal *identity* rather than addressing the *politics* of difference that structure the fuel chain.

The NWMO produces difference in its study by creating discursive and material distinctions between Aboriginal peoples and others. Aboriginal knowledge, for example, is clearly differentiated from scientific, Western, and expert claims in terms of its power to rationalize, its breadth of application, and its relevance. Ironically, Aboriginal difference serves to universalize the industry's positions and claims – representing them as normal and those of Aboriginal peoples' as special: spiritual, cultural, different – and to position Aboriginal knowledge very differently in relation to the production of knowledge about nuclear fuel waste. This difference is also a red herring: it obscures the ways in which many Aboriginal peoples are positioned within the fuel chain and so ensures that the structures that systematically and systemically unload the negative effects of nuclear production onto Aboriginal peoples' lands and livelihoods go unexamined. As Sharma (2006, 28) notes, official diversity, the kind often embraced by the state, perpetuates oppression: "It legitimates the continued organization of difference in order to both organize and legitimate the subordination that the differentiated experience." Indeed, the NWMO's management of diversity produces conditions that Young calls the paradox of cultural imperialism: being marked out as different while being invisible at the same time (1990, 58-61).

In closing, I would like to make the following observations about the normative dimensions of this policy-making process. It is in the nuclear industry's interest to keep certain relationships between people and the nuclear fuel chain hidden. This is the case with the relationships between nuclear production and many Aboriginal peoples such as the SRFN. The experiences of many First Nations such as the SRFN challenge both the content and the method of the nuclear industry's claims about radioactive materials, their effects, their social and environmental legacies, and their appropriateness for Canada and therefore trouble the industry's narratives about the safety and appropriate management of nuclear fuel waste and history of the nuclear fuel chain. Because the fuel chain disproportionately affects the lands and livelihoods of Aboriginal peoples, particularly First Nations, they are an important source of knowledge and critique. Their experiences are also the basis of some of the few counternarratives to those of the nuclear industry about the effects of radiation and therefore are unique in that they constitute a distinct threat to the assumptions, knowledge, and goals of the nuclear industry with respect to the fuel chain.

The future of the nuclear program in Canada, along with very powerful sets of interests, is at stake in the management of nuclear fuel waste. The powerful position and seemingly unlimited agency of the industry are, however, fragile. As suggested in this chapter, maintaining them requires continual work and protection. The dominance and power of the nuclear industry and the NWMO's discourses are reliant on the disqualification by various means of the knowledges of Aboriginal peoples (as well as others). Some of these means I have described above. The policy discourses of the NWMO and the nuclear industry are dominant and constitute the official knowledge on this issue because they successfully marginalize the claims of Aboriginal peoples and obscure from view their experiences and geographies. Paradoxically, the incorporation of Aboriginal peoples into nuclear fuel waste management signals more preservation of the status quo and universalization of the assumptions and logics of the nuclear industry than recognition of the alternative knowledges and experiences of Aboriginal peoples. This is, sadly, a status quo that ensures that the lands and livelihoods of Aboriginal peoples will continue to be negatively implicated in the political economy of the fuel chain.

Acknowledgments

I am grateful to D. Memee Lavall-Harvard, president of the Ontario Native Women's Association, for her comments on an earlier version of this chapter and to the editors of this volume for their helpful suggestions. I would also like to thank Lillian Trapper and Dr. Patricia Fleming for their insightful reviews.

Notes

1 The term "Aboriginal peoples" is an umbrella designation that refers to three distinct group-ings of Indigenous peoples recognized by the Canadian Constitution: First Nation, Inuit, and Métis. All three groups have distinct historical, territorial, and political relationships with the crown, and all three include diverse political, linguistic, cultural, and geographical groups and nations. Recognizing the diversity within and between Aboriginal peoples, I use the term as a shorthand to refer to more than one of the three distinct groupings at the same time and, where necessary, to refer to the use of the term by the Nuclear Waste Man-agement Organization (NWMO). The NWMO has been criticized by groups, such as the Assembly of First Nations (AFN), for using a pan-Aboriginal approach and incorporating "Aboriginal peoples" as an undifferentiated and singular category into its study. See, for example, AFN (2005c).

2 Unfortunately, there are no independent or verbatim records or transcripts of the NWMO's consultations with Aboriginal peoples. I relied therefore on transcripts of testimony given by First Nations communities such as the Serpent River First Nation at public hearings about similar nuclear issues, including hearings into the deep geological disposal of nuclear fuel waste, radioactive tailings disposal, and issues around uranium mining, and the submissions presented by Aboriginal groups to the NWMO. I am also aware that the claims I am making about these knowledges are partial: they are my interpretations, based on my understand-ings of the documents consulted, and certainly are not neutral. I am not an Aboriginal person, and my understandings are politically situated.

3 I do not want to suggest by this that the formulation of Aboriginal peoples' contributions as values, perspectives, and insights (as shown in Figures 7.1 and 7.2) in the ten questions is not in itself problematic. Indeed, in other work (Stanley 2006, 2008), I have problematized this specific formulation, as have Aboriginal organizations involved in the NWMO process (e.g., AFN 2005e).

4 "Canadian" is an imagined homogeneous identity. Although it appears as a general (nation-ally inclusive) category, it is in fact a specific representation of a particular – and perhaps fictive – experience (Stanley 2008). On the concept of the national imaginary, see Anderson (1983); see also Bauder (2006); and Sharma (2006).

5 I am unaware, at the time of writing this, of alternative Aboriginal interpretations of this teaching. I make this distinction to respect the diversity between Aboriginal experiences and practices and epistemic and moral concepts. The AFN's and SRFN's interpretations of the seven generations teaching may differ, in ways of which I am unaware, from those of other First Nations or Aboriginal peoples.

6 The AFN is a national umbrella group lobbying on behalf of all First Nations in Canada. Its mandate descends from a national assembly of First Nations chiefs. The AFN's mandate is to lobby the government on behalf of all First Nations in Canada and to advance the rights of First Nations in Canada in ways consistent with the mandates of the National Assembly. The AFN does not have the political, moral, or constitutional ability or mandate to negoti-ate or consult on behalf of First Nations. With respect to its participation in the NWMO's process, the AFN remained constant in its message that its participation constituted not "consultation" but dialogue and that the NWMO, to fulfill its mandate, needed to consult with individual First Nations.

8

Canadian Communities and the Management of Nuclear Fuel Waste

Brenda L. Murphy

The management of nuclear fuel waste in Canada is at a crossroads. Following decades of concept-oriented, technological planning, the recommended approach of the Nuclear Waste Management Organization (NWMO) for long-term waste management will begin to "take place" sometime within the next several years. The emplacement of Adaptive Phased Management (APM; see Durant and Stanley, this volume) is not expected to begin for several years and possibly longer because the NWMO intends to undertake an extensive public consultation program prior to commencement of the siting process. This chapter uses insights gained from Canadian and Swedish case studies drawn from all segments of the nuclear fuel chain to extrapolate the potential future effects on Canadian communities as decision making begins to take place.

This chapter focuses on geographically peripheral communities in Canada's North, and on under-resourced non-governmental organizations (NGOs), since it is most likely in these spaces that a long-term management facility will be constructed and since it is these social groups that often provide the critique of the dominant nuclear establishment perspective. The cases include Canadian uranium mining and milling, the review of the deep geological disposal concept by the Seaborn panel, the management of historic low-level waste in Port Hope, Ontario, and Sweden's nuclear fuel waste program.

To date, Canada's approach to nuclear fuel waste management has been placeless, focused on technical concepts rather than emplaced, site-specific decision making. Emplaced decision making, as undertaken by Sweden and many other countries (see Murphy and Kuhn, this volume), involves the evaluation of nuclear fuel waste management strategies in which the assessment of technical options is tied to particular sites. This approach allows for a real-world evaluation of the potential impacts of a proposed project. In Canada, until APM was accepted, there were essentially two limitations on the possible locations for long-term nuclear waste facilities. Deep geological disposal was limited to the vast Canadian Shield in northern Canada, and on-site storage was restricted to the reactor sites in Ontario, Quebec, and New Brunswick. The third option, centralized storage, was not associated with any locational restrictions other than the boundaries of Canadian territory. The NWMO's 2005 final recommendation, accepted by the Canadian government in June 2007, removed the first territorial limitation by suggesting that both the Canadian Shield and areas underlain by Ordovician sedimentary rock (in southern Canada) could be considered for deep disposal facilities. The only limitation imposed by the NWMO's final recommendation was that any developed facility should be sited within Ontario, Saskatchewan, Quebec, or New Brunswick: that is, in provinces that have benefited from the employment and electricity associated with the nuclear industry. This is a vast territory within which is located a multitude of communities that have been, will continue to be, or may be affected by Canada's nuclear policies and practices.

I raise three key arguments in this chapter. First, just as the impacts of past radiological events can serve as an analogue for potential future impacts, so too can the past impacts of nuclear policies and approaches be used to understand the potential impacts, both positive and negative, on Canadian communities. The Canadian nuclear industry's legacy related to uranium mining and early nuclear power and waste management approaches, including community radioactive contamination and unilateral decision making, is likely to generate animosity and conflict, as well as ardent support, during any new siting exercise. Second, there have been multiple repercussions of the policy of placelessness, including the inability to engage local level stakeholders in the nuclear fuel waste policy assessment exercises, concerns over the representation of "public" perspectives by NGOs, and the domination of policy and management approaches by the proponent and other government agencies. Third, as Canada moves into the next phase of nuclear fuel waste management and as policies begin to take place, several additional problems are expected to emerge. Underpinned by the historical context, with more tangible, place-specific data, both geographic and interest-based communities will be able to evaluate the proposed project's impacts and will become further politicized either to support or to oppose the proposed facility. Such an evaluation will lead to schisms and tensions within and between communities. Furthermore, it appears that avoidance of the siting question is currently hampering geographic communities from participating fully in the development of nuclear fuel waste management strategies. However, once siting begins, these local players may exert latent influences on the process; they may be able either to derail or to push through the proposed facility, depending on their levels of support or opposition.

This chapter extends Ulrich Beck's (1992, 2005) idea of the subpolitics that exists within a "risk society" through the consideration of issues associated with communities. Since this chapter is concerned with effects of place and placelessness, it differentiates between two groups involved within the subpolitics: those communities rooted in place, geographic communities, and those whose affiliations are related to common interests or worldviews, which I label interest-based communities. The chapter also incorporates a temporal aspect by exploring how communities are "stretched out over time." After outlining the theoretical orientation and examining case studies, I offer some recommendations for what could be done to ameliorate these negative consequences.

Community and Modern Risk

According to Beck (1992, 1995), as a society focuses more on risk, it becomes centred on a "negative utopia" and the logic of disposition and uncertainty. In a risk society, conflicts related to the distribution of "bads" (dangers and externalities) either replace or are superimposed onto traditional conflicts. Conflicts "erupt over how the risks accompanying goods production ... can be distributed, prevented, controlled, legitimised" (Beck 1994, 6). Widespread acknowledgment of these risks, particularly within civil society, tends to undermine the legitimacy of both technoscientific and government actors. When various communities and groups begin to apprehend and experience the dangerous side effects associated with modern technologies, trust in government authority and scientific methods and promises is undermined. Narrow, technocratic definitions of danger, hazard, risk, and externality are challenged by lay actors who develop broader, more contextualized definitions of risk and increasingly demand a voice in decision-making processes that affect their lives and landscapes (Beck 1994, 1995). Consequently, decision-making processes become ever more complex and fractious, with the range of community voices vying to be heard greatly increasing (Giddens 1984, 1994). Such decision-making processes provide a space within which subpolitics can flourish, including the activation of citizen groups and the development of social movements that have a plethora of contradictory agendas often challenging the status quo (Slovic 1987). As Beck writes, "the themes of the future [ecological salvation, renewal of the world] ... have not originated from the farsightedness of the rulers ... and certainly not from the cathedrals of power in business, science and the states ... They have been put on the social agenda ... by entangled, moralizing groups and splinter groups" (1994, 19).

Discourse within these spaces of subpolitics calls for increasingly pluralistic decision making in which legitimate, meaningful mechanisms for participation are provided (see, e.g., Oliga 1996; and Reddy 1996). From a civil society perspective, the imperative is that all communities should be able to participate in decision-making processes that affect their lives (Fiorino 1990). In contrast, the tendency of government authorities and the technoscientific community is to circumscribe risk and truncate the decision-making process

(Jacob 1990). Rees (1985) suggests several ways in which those in authority attempt to control the decision-making process. Two of them include focusing on local facility siting issues rather than questions of national policy and capturing problem definition by framing the issue in technical terms.

Although Beck identifies the realm and importance of subpolitics, beyond broad ideas associated with gender and class, he has much less to say regarding the composition and identity of community actors within the civil society sphere who are clamouring for inclusion in decision-making processes. Although he asserts that traditional forms of community are beginning to disappear and that newly formed social relationships and networks "now have to be individually chosen" (1992, 97), at the scale of unravelling the dynamics of decision-making processes, he does not provide insight into the nature of those relationships and networks, nor does he differentiate among actors according to their affiliations. Since decision making within a risk society will have tangible but differentiated impacts on all communities, analysis of these impacts must recognize the groups of actors involved and evaluate both their capacity to influence processes and policies and the ultimate outcomes. Furthermore, since risk-related policy decisions that involve the construction of noxious or hazardous waste facilities often disproportionately affect localized, peripheral geographic areas (see Bullard 1999), it seems appropriate to distinguish between communities tied to particular places and those related by interests or worldviews.[1]

Place-specific, geographically bounded communities are constructed through face-to-face contact. They include local protest groups, neighbourhoods, and municipalities (Miller 1992, 31). Interest-based communities may or may not have a geographic reference point. Instead, they are often "stretched out" over space, have no clearly defined boundaries, and sometimes exist only in the virtual world (Silk 1999). These types of communities develop through the bonds of common interests, worldviews, or kinship networks. Examples include the nuclear advocacy community, religious communities, and environmental NGOs. In this chapter, a secondary distinction is also made regarding the temporal dimension of communities – both geographic and interest-based communities have relationships that extend backward and forward through time, incorporating the idea of sustainable development. On the one hand, the incorporation of historical contexts helps to situate and ground contemporary problems and issues. On the other, the temporal dimension provides an opportunity to move assessments beyond the present into an evaluation of impacts on future generations.

This categorization, however, does not imply that communities are static or mutually exclusive. It is acknowledged that communities are imagined, fluid, social constructs with overlapping boundaries and evolving memberships (Silk 1999; Young 1990). For instance, within a geographic community, such as a municipality, there could also be neighbourhood-level (e.g., a neighbourhood association) and interest-based communities (e.g., a religious group or service organization) whose boundaries might transcend or are not coterminus with those of the municipality. Another example would be the mapping of municipal borders over territory previously organized and delineated by Aboriginal peoples.

As the NWMO rightly notes, "community is not readily defined along geographic or political boundaries. A community may reflect shared perceptions and attitudes, and shared socio-economic foundations. It may be defined in part by behaviour patterns which individuals or groups of individuals hold in common, through their daily social interactions, the use of local facilities, participation in local organizations, and by involvement in activities that satisfy the population's economic and social needs" (2005c, 275).

In terms of decision-making processes, it is important to understand that different communities establish their own sets of moral norms and mores and that they affect group understandings of risk and its acceptability. As Douglas writes, "as soon as there is a community, the norms of acceptability are debated and socially established. This activity constitutes the definitional basis of community ... A community uses its shared, accumulated experience to determine which foreseeable losses are most probable, which probable losses will be most harmful, and which harms may be preventable. A community also sets up the actor's model of the world and its scale of values by which different consequences are reckoned grave or trivial" (1991, 180).

In the following sections, I use these ideas about community and the risk society to underpin the discussion of Canada's and Sweden's nuclear fuel waste management programs. This discussion utilizes the impacts of past nuclear policies on communities as both the historical context that underpins contemporary and future nuclear fuel waste undertakings and an analogue of the potential future impacts as nuclear fuel waste policies begin to "take place," as opposed to being placeless.

Placing Nuclear Waste Management: Historic Distribution on the Canadian Landscape

Although Canada has yet to site a facility for the long-term management of nuclear fuel waste, the nuclear industry has a long and complicated past. Its activities have had many impacts, both positive and negative, on a host of geographic communities, many in peripheral areas (e.g., northern, rural, and marginalized regions without substantial social-political access to power and resources). Furthermore, the vast majority of facilities were developed on land claimed by various First Nations and Métis peoples – who are among the most marginalized of Canadian peoples. This history, beginning with the mining of uranium, provides the context for future nuclear fuel waste undertakings since it will profoundly influence the nature of upcoming interactions with the NWMO and other nuclear authorities. This history can also serve as an analogue regarding the potential future community consequences associated with siting attempts or other management initiatives.

Many geographic communities, often located in Canada's periphery, such as Kincardine, Deep River, and Blind River, Ontario, generally support the nuclear development in their areas (although there are dissenters on the margins) since the facilities are associated with positive outcomes such as much-needed employment opportunities. For instance, Blind River's mayor is strongly in favour of building a new nuclear reactor along the north shore of Lake Huron and has been trying to solicit support from neighbouring communities (Northwatch 2006). Similarly, Al-Haydari's (2007) research indicates that in Kincardine, local government officials and geographic community representatives, such as Kincardine's Chamber of Commerce, strongly support the nuclear industry and the proposed deep geological facility for low and intermediate nuclear waste. This is a common pattern among peripheral geographic communities hosting various types of nuclear facilities (see Blowers 1999 for several examples). For instance, as Murphy and Kuhn (2006) note, due to the economic benefits of a military nuclear waste facility, the municipality of Carlsbad, New Mexico, again located in a rural area with few other economic opportunities, solicited and adamantly continues to support the deep geological repository, with little local opposition.

Along with such positive benefits and support for the nuclear industry, there have been significant negative impacts on various geographic communities, including Aboriginal peoples, across Canada. Canada's nuclear industry began in the 1920s and 1930s, with some of the earliest detrimental effects associated with the mining of uranium in Port Radium, in a remote part of the Northwest Territories (NWT), and ore processing in Port Hope, Ontario (a small eastern Ontario municipality). Subsequent uranium mines were also opened at Rayrock, NWT, Uranium City, Saskatchewan, and near Bancroft and Serpent River First Nation in Ontario, all areas remote from centres of power and decision making in south-central Ontario. Much of the uranium for the American Manhattan Project and Cold War initiatives was provided by these early mines (Blow 1999). Within Serpent River First Nation's territory, the community of Elliot Lake was developed to house the miners and their families. Today all of these mines are closed, and many are undergoing decommissioning.

All of these sites experienced some level of contamination from uranium tailings and processing. During the 1950s and 1960s at Elliot Lake, "routine effluent discharges, seepage and containment failures combined with generally poor management practices, resulted in an inacceptable [sic] level of contamination of the Serpent River System" (Energy, Mines, and Resources [EMR] Canada 1982, 191). In Ontario and Saskatchewan alone, there are over 200 million tons of uranium mine tailings to be managed (Edwards and Del Tredici 1999, 3). At Port Hope, the processing facility is still in operation, with the community struggling to deal with a historic low-level nuclear waste problem. The Auditor General of Canada report (1995, 3-10) stated that there are 1 million cubic metres of waste to be managed. In the Northwest Territories, the communities have experienced such high levels of morbidity and mortality that Deline, a Dene settlement, has been dubbed "The Village of Widows" (Blow 1999).

Today uranium mining occurs exclusively in the Athabasca Basin in northern Saskatchewan, providing approximately 40 percent of the world's uranium. The consequences of these newer mines remain unclear. Some observers, such as Mining Watch Canada, maintain that there have been negative consequences associated with radioactive materials leaking into the surrounding watershed (see Kneen 2006). Other observers, such as industry analysts, insist that all negative impacts have been mitigated (Saskatchewan Mining Association). Given both the employment provided by the mines and the various other programs supported by the industry, plus the uncertainty associated with the long-term environmental impacts and the uranium industry's past track record, it is not surprising that the Cree and Dene/Chipewyan peoples living in the area offer mixed support for the mining operation (Parsons and Barsi 2001).

Other geographic communities that have been affected by the nuclear industry include Deep River and Chalk River, Ontario. Both towns are located in rural areas but have for some time been the home of Atomic Energy of Canada Limited's (AECL) nuclear research and development laboratories. These laboratories, located on Algonquin territory, provide many well-paid, professional jobs. Various forms of nuclear waste are also stored on site. These are the externalities from the various development projects. Chalk River and Deep River were both involved in a scheme that would have seen the cleanup of the wastes associated with the Port Hope facility and the laboratory. That initiative was scuttled in 1997 by the federal government (Murphy 2001).

Today, in addition to the factory at Port Hope, different aspects of uranium processing also occur in Peterborough and Blind River, Ontario. The latter facility is located two kilometres from the Mississauga First Nation reserve, on land considered part of its territory. Although information about the impacts of these facilities is not easily obtained, it appears that the usual differences of opinion between industry and opposition groups prevail. In the case of the Blind River facility, a recent approval to expand operations has caused "shudders" among one group concerned with the uranium-laced fumes being emitted from the facility's incinerator (International Institute of Concern for Public Health [IICPH] 2007). In contrast, the industry insists that the environmental impacts are negligible.

In terms of nuclear fuel waste, other than small quantities at the Chalk River laboratory site, virtually all of the waste is stored at the nuclear power plants. There are three sites in Ontario, at Pickering, Darlington, and Kincardine, and one site each in Quebec and New Brunswick. Although both Pickering and Darlington are located in south-central Ontario, the remainder reflect more marginal locations. Ninety-five percent of the waste is produced and managed by the Ontario plants. The Kincardine facility, now called Bruce Power, was sited within the territory of Saugeen and Nawash First Nations. The Aboriginal communities were not consulted when the Bruce plant was built, and later, in the 1990s, when the facility owners applied for an expansion, the Aboriginal communities were unable to trigger a full-panel environmental assessment despite the earlier siting infringement (Murphy 2001). More recently, although draft environmental guidelines for the proposed Kincardine facility include specific reference to potentially affected First Nations communities, the Métis Nation of Ontario (2008, 12) maintains that no mention is made of effects on Ontario's Métis communities.

Initial attempts to undertake scientific research into suitable rock formations and site-specific information for a long-term nuclear fuel waste disposal facility began in the mid-1970s. Subsequent to vigorous protests at places in rural and eastern Ontario, such as Massey and Madoc, the federal government chose to develop its underground research laboratory and establish the community of Pinawa in Manitoba – with the express caveat that no nuclear waste would ever be buried at this site (Brisco 1988). Today the facility has been substantially mothballed. It is important to emphasize here that these early protests are considered to be one of the key reasons that Canadian nuclear fuel waste policy became focused on a placeless, concept-level approach and that the underground facility was sited outside the province that produces the most waste (Murphy 2001).

This brief background of the Canadian nuclear chain, from mining to electricity and waste production, demonstrates that there has been and continues to be support for and positive benefits associated with the nuclear industry and its developments. However, there has also been a legacy of externalities associated with radioactive contamination, especially in peripheral geographic communities, including those of many First Nations.

I contend that this historic legacy will underpin both the positive and the negative perspectives and capacities encountered by consultation initiatives or siting attempts undertaken by the nuclear industry or the NWMO. In essence, this past legacy, the current situation, and the future contexts are associated with the temporal dimension. The impacts and perspectives will affect communities as they "stretch out over time." By way of example, since the federal government has accepted the NMWO's final report (2005c), both Ordovician (southern Ontario) and Canadian Shield rock (northern Ontario) are deemed eligible mediums for a deep geological disposal facility, whereas previously only the latter was considered suitable. The NWMO will now be able to include municipalities, such as Kincardine, that already have a positive relationship with the agency and will not have to rely on the more remote areas that have experienced the worst of the externalities associated with the nuclear industry (see also Chapter 9, this volume).

It is also clear that for the peripheral geographic communities, the nuclear legacy has had both positive and negative impacts on their current and future sustainability and capacity. In some cases, the nuclear facilities have meant employment, but there has also been infringement on Aboriginal territorial rights, conflict over nuclear management strategies, and soil and water contamination. At a broader scale, this historical context provides the background for understanding the positioning of the placelessness policy within nuclear waste management and underpins what will occur as nuclear policies begin to take place.

The Impacts of "Placelessness" and "Taking Place"

The Canadian Context

This section focuses on nuclear fuel waste policy approaches and processes, including the Seaborn panel's review of environmental assessment of the nuclear fuel waste disposal concept (Canadian Environmental Assessment Agency [CEAA] 1998). It assesses the impact of the nuclear policy of placelessness and postulates the effects on communities of future undertakings as policies begin to take place. As the NWMO undertakes new processes, two trends can be predicted. First, there will be a cost to communities (e.g., burnout, increased community tension, etc). Second, although geographic communities have not played a significant role up to the present time, their latent power and influence will become increasingly evident as the NWMO's plans begin to take place. Additional insights into potential impacts and the role of geographic communities are gained from examining the situation in Sweden because the siting process is far more advanced in that country. Other case studies, such as Port Hope, Ontario (for historic low-level waste), are also incorporated to flesh out the picture.

The Impacts of Placelessness

Critiques of the placelessness policy suggest that it has been a deliberate tactic of both the nuclear industry and the government to avoid the controversy associated with the identification of particular sites. Indeed, the mayors of five large northern Ontario municipalities have expressed opposition to the siting of a nuclear fuel waste facility within their jurisdictions, and some have passed official resolutions to that effect (Northwatch 2004; see also Murphy 2001 and www.nukewaste.ca). Industry supporters maintain that separating technology from place is simply a logical approach, allowing the rational determination of the management approach prior to engaging in siting exercises (NWMO 2005b, 145-46, 232, 430). Regardless of the veracity of either viewpoint, the reality is that, for struggling, remote communities, when confronted by a concept-level review of a future technological undertaking that may or may not affect them, it is difficult to stimulate interest. This was noted by the Seaborn panel (CEAA 1998) in its review of the deep geological disposal concept (see Murphy 2001). For geographic community perspectives, the input of those currently dealing with other aspects of the nuclear chain has been the primary way to gain insight into the possible views of a siting-affected community as well as the potential impacts of a long-term nuclear fuel waste management facility. Within Canada, such geographic communities include Port Hope (for historic low-level waste); Serpent River First Nation and Elliot Lake (for uranium mine tailings); First Nations communities of Saugeen and Nawash; Kincardine, Pickering, and Darlington (for short-term storage of nuclear fuel waste); and Deep River/ Chalk River (for the failed low-level historic waste siting process and proximity to the AECL laboratory).

In Canada, the involvement of various communities in policy debates has also been limited by the "nuclear establishment," which has been described as extremely "concentrated, bureaucratic, and inaccessible" (Mehta 2003). For instance, research clearly showed that government and industry officials predominated throughout the process of developing the terms of reference for the Seaborn panel. Other communities, such as environmental NGOs, were provided only with token, belated involvement. No information was available, nor was any public consultation provided, until after the terms of the environmental assessment were substantially set (Murphy and Kuhn 2001). Echoing Jacob (1990), Mehta (2003) outlines that the nuclear establishment consists of a coalition of regional jurisdictions (e.g., provincial electricity generators), a national coalition related to industrial interests (e.g., Canadian Nuclear Association [CNA], AECL), the national-level bureaucracy (e.g., Canadian Nuclear Safety Commission [CNSC]), and supportive Senate and House committees in Parliament. Mehta argues that the former Atomic Energy Control Board (AECB; now CNSC) was given a broader range of powers, with less oversight and concern for public participation than were other federal regulatory agencies, due to the nuclear industry's genesis through the Manhattan Project, followed by the security concerns associated with the Cold War. Mehta contends that concern about the closed nature of the nuclear establishment has contributed to the formation of numerous local and regional NGOs, interest-based communities, such as Durham Nuclear Awareness, Northwatch, and the Inverhuron Ratepayers Association, as well as national groups such as Greenpeace and Energy Probe.

I contend that one way in which the nuclear establishment has been able to maintain hegemony has been through the policy of placelessness. Without a specific site and set of impacts, geographic communities have struggled to understand the potential impacts of any new policies and concepts. They have had to rely on knowledge gleaned from past and current contexts. This policy of placelessness has also made it difficult for First Nations, Inuit, and Métis communities to evaluate nuclear initiatives since their traditional knowledge is by definition place based and contextual. Consequently, a substantial part of the non-Aboriginal appraisal of Canadian nuclear policies and processes has been orchestrated by interest-based groups whose concerns are less rooted in place, particularly the regional and national NGOs and religious organizations (for more on this latter contribution, see Timmerman, this volume).

Interestingly, during the Seaborn panel's review of the deep geological disposal concept, a discourse developed regarding the nature of the "public" and the extent to which these interest-based groups represented it. Despite the genesis of many of these interest-based communities as a direct challenge to the nuclear establishment, members of the predominant nuclear establishment often contended that the interest-based communities were merely "special interest groups" and not representative of the "public" (Murphy 2001). The NGOs, of course, disagreed and asserted that their regional and national organizations often acted on behalf of many small local groups and others who had neither the time nor the expertise to plough through the volumes of technical data associated with the environmental assessment. Even Dr. F.K. Hare, who had chaired an inquiry into nuclear waste management in the late 1970s, waded into the debate. Hare (1998) declared that the Seaborn panel had only heard the views of a highly selective group and that widespread public support was impossible. Although one can argue that the NGOs did not represent everyone, to suggest that they represent "only" special interests is simply a deliberate tactic to undermine their legitimacy. This undermining effort must be seen in the context of a stinging NGO critique that, ultimately, profoundly influenced the Seaborn panel's review (see Chapter 3, this volume). I maintain that, as the NWMO moves into future processes, the issue of what constitutes "the public" will continue to be problematic. Even as siting takes place, little wide-scale public participation can be imagined beyond the locally affected area. Thus, NGO participants will continue to provide a perspective that, even if it is not representative of some nebulous public, at least provides a salient critique to the views of the predominant nuclear establishment.

To counter the criticism that the public has not been widely consulted on nuclear policy, recent consultation processes conducted by the NWMO have been careful to include a wide range of pre-established geographic and interest-based communities as well as soliciting the views from all of Canada's Aboriginal groups and non-affiliated members of the public (see www.nwmo. ca). However, the reporting of those consultations has been opaque, with contrasting perspectives typically unaccredited and simply summarized by ambiguous categories such as "some people said" and "others suggested" (see, e.g., NWMO 2004e). This ambiguity has served to undercut, dilute, and sanitize the arguments being raised by both knowledgeable geographic communities as well as the interest-based NGO communities. Furthermore, although the NWMO has indeed consulted widely, it is uncertain to what extent those opinions have had any real influence, since the long-term strategy recently recommended by the NWMO and accepted by the federal government (e.g., phased deep geological disposal) varies little from that put forward by the nuclear establishment in the 1980s and 1990s (see Chapter 9, this volume).

Impacts of "Taking Place"

In the future, once the preferred strategy is finalized and the process of emplaced policy decisions begins, the public representation issues described above will become ever more salient and contentious. Controversy regarding the role of regional- and national-scale interest-based communities in a "local" process will have to be addressed, conflict tied to differences in worldviews will emerge among communities, and there will be real, tangible impacts on community members.

Members of both interest-based and geographic communities maintain that, whenever they are involved in facility siting even at preliminary stages, or in other issues in which strong opinions are expressed, these processes divide communities and make things "awkward" (Murphy 2001). For instance, during the review of the historic low-level waste facility for Port Hope, some community members found it distressing to spend all day in "difficult dealings" with individuals and then ending up at the same school picking up their kids. It was also noted that there were sometimes clashes between the Community Liaison Group (i.e., members of the public selected to review the proposal) and members of "special interest groups." A third concern raised was related to the perceived dominance of interest groups at public consultation exercises. It was suggested that people were tired of competing with these dominant voices and felt threatened if they spoke out in favour of the project.[2] Notice among these concerns the underlying tension within the geographic community as well as between the geographic and interest-based communities.

The interest-based NGOs noted that sitting in hearings day after day for the review of the deep geological concept was both exhausting and time consuming (Murphy 2001). NGO representatives found the animosity toward them at these hearings emotionally draining. As one stated, "it takes its physical and emotional toll to be in a room day after day with people who disdain you and who disrespect you at a personal and political and professional level" (cited in Murphy 2001, 153). These actors also commented that they felt further marginalized during the hearings because of the proportions of proponent versus public and men versus women. "We used to do both gender counts and proponent-public counts ... The ratios were usually running in the neighbourhood of ten to one, sometimes eight to one, sometimes twelve to one, both in terms of the gender ratio but also in terms of the proponent" (cited in Murphy 2001, 206). Although these are only brief examples, they hint at the range and depth of the impacts that any future nuclear fuel waste undertakings will have on communities.

My purpose here has been to examine current contexts and past processes within the Canadian nuclear fuel chain to gain some insight into the past effects of the placelessness policy and what might happen as emplaced policies develop. However, much is yet unknown. To further understanding of the possible future impacts as policies take place, I present below a brief Swedish case study. The point here is not to fully explicate this example but to look for insights applicable to the Canadian case.[3] Since in the Canadian case geographic communities have been marginalized by the placelessness policy, the focus of this section is on the role of the place-based communities within emplaced decision making as well as their relationships with interest-based communities and the nuclear establishment.

The Swedish Context

The Swedish approach to nuclear fuel waste management, particularly as articulated by the geographic community of Oskarshamn, is considered to be one of the most fair, inclusive, and democratic in the world. It is hence upheld as a model for other countries to emulate. I suggest that it is particularly useful in modelling the involvement of geographic communities but has continued to struggle with the development of fair processes for interest-based communities. One broad contextual factor is important to note before outlining the case study – Sweden has, at least in theory, decided to phase out nuclear power. Although the termination date is far from certain, the commitment to phase out has contributed to the willingness of some Swedish communities to work with the authorities to find a solution to the safe management of existing waste.

Oskarshamn has a population of approximately 27,000 and encompasses a territory of about 1,000 square kilometres. It is located in a rural area, several hours from Sweden's economic centre, Stockholm. The first nuclear reactor was established there in 1972, about twenty kilometres northeast of the town on the Simpevarp peninsula. As is the case with Canada's Kincardine facility, as well as the uranium mines in northern Saskatchewan, the nuclear facilities in Oskarshamn provide significant local employment benefits. Although the current economy in Oskarshamn is strong, with low unemployment, it is dependent on a few large industries, including the nuclear complex (involving three power plants), the nuclear fuel waste interim storage facility, the Aspo deep geological laboratory, and the research facility associated with the proposed nuclear fuel waste encapsulation plant. About 1,000 jobs are directly generated by the nuclear industry, and the power plants contribute about 10 percent of Sweden's electricity supply (Oskarshamn 2003).

In the 1990s, Svensk Kärnbränslehantering AB (SKB) (i.e., the proponent, equivalent to the NWMO) approached the community of Oskarshamn to undertake a nuclear fuel waste geological disposal feasibility study. SKB had not been successful in finding a potential host community in an open call for volunteers and thus approached those communities with already established ties to the industry (Sundqvist 2002). Of importance here is a significant point of difference between Canadian and Swedish municipalities. Canadian municipalities do not have independent decision-making authority; instead, they have their responsibilities delegated to them by the provinces. Swedish municipalities have independent authority, guided by federal legislation and policies, over most affairs internal to their borders. Thus, due to their inherent capacity to manage and control local land use decisions, plus their extensive involvement in nuclear issues, the geographic community of Oskarshamn was strongly positioned to make several key demands of the national government. The municipality requested (1) an integrated review of the entire final nuclear fuel waste disposal system, including encapsulation, transportation, and deep repository, and (2) the establishment of clear national policies for the waste management system, prior to specific site-level decision making (Sundqvist 2002). It demanded recognition that nuclear waste was not merely a "local" problem and that municipalities should play a role in setting national policy. Due to pressure from the municipalities, particularly Oskarshamn, legislation was passed requiring the provision of an annual grant, financed through the national Nuclear Waste Fund, to all municipalities involved in feasibility studies (Community Waste Management (COWAM) 2002-3). They were also successful in extracting a guarantee that the national government would not use its veto to override any decision made within the legitimate jurisdiction of the local authority.

Following from extensive local consultation processes, the municipality developed the "Oskarshamn model" to structure participation in the encapsulation environmental assessment and the feasibility study. This set of principles encompasses ideas such as the need for openness and effective participation, the importance of the municipal council as the legal representative of the local community in a democracy, and the need to involve the public and NGOs (Oskarshamn 2003).[4]

Following completion of the feasibility study in 2002, and over a year of debates and initiatives within the municipality, Oskarshamn agreed to allow SKB to move on to the site characterization study with the proviso that thirteen conditions be addressed. Among these conditions were the requirements that Oskarshamn and other communities undergoing similar studies receive guaranteed financing, that any facility would only take Swedish nuclear fuel waste, and that a more open, co-ordinated approach to safety was needed, acknowledging the potential for the compounding effects of the facilities within the Oskarshamn nuclear complex (Oskarshamn 2003). Monitoring of the fulfillment of these conditions is being undertaken by six working groups involving politicians, association members, environmental groups, and ordinary citizens.

Despite the comprehensive, proactive approach developed by the geographic community of Oskarshamn, the interest-based communities have several misgivings about Sweden's nuclear policies and have felt marginalized by the process. No fewer than seventeen "rescue groups" have sprung up to protest Swedish policies and approaches as well as to question the strategies put forward by the municipalities. Since they have had little funding support, the interest-based groups formed a loose coalition called the "Waste Network" that co-ordinates groups of citizens in areas considered as possible waste management sites (Holmstrand 1999). This is similar to the Canadian group Northwatch, which was developed prior to the Seaborn panel's review process as an umbrella organization representing many small protest groups and environmental organizations from northern Ontario (Murphy 2001; see also Chapter 3, this volume, and www.nukewaste.ca). The goal of the Waste Network is to promote the exchange of experience and knowledge among local groups and to co-ordinate responses and actions to agency reports and dialogues. The Waste Network's key viewpoints include that, (1) since the problem of nuclear waste has not been solved, the further operation of reactors should be restricted to the extent possible, (2) an independent authority, not SKB, should supervise the management process, and (3) environmental organizations should be viewed as legitimate representatives of the public and should be given "reasonable conditions and resources" to take part in environmental assessment processes (Holmstrand 1999).[5]

Holmstrand further asserts that, although nuclear fuel waste management is a national issue, "siting in Sweden has never been more than a local question" since broader policy contexts regarding the choice of methods have never been subjected to a democratic debate but have been defined as narrow technical issues (1999). Furthermore, as WISE Sweden (a Swedish NGO network) insists, SKB has targeted its siting efforts on municipalities with weak or non-existent opposition – for instance, regions with high unemployment and/or areas reliant on the nuclear industry for jobs and prosperity (Tornqvist 1999).[6]

Moreover, although Oskarshamn has been able to arrange both a national participation venue and independent funding for all municipalities involved in the siting process, until recently similar concessions for non-place-based actors, such as NGOs, were not forthcoming. Although it was possible for NGOs to obtain some funding, it was sporadic and at the discretion of the players involved, including the municipalities, which could elect to share some of their funds with NGOs (e.g., the protest group SOS Tierp received some funding from the municipality). Recently, legislation has been passed that provides for a somewhat more equitable allocation of financial resources to NGOs. Financial support is limited to national-level organizations with large memberships, even though this is partially overcome by allowing umbrella or coalition groups to apply for joint funding. However, despite these changes, the funding rules still leave many interest-based groups without financial resources (Sundqvist 2007, personal communication).

Holmstrand (2003) has noted that there was a polarization of perspectives between the "establishment and the NGOs," resulting in the lack of any type of constructive relationship between the two. At least from the perspective of the NGOs, the proponent did not appear to trust citizen groups and NGOs. Although still far from perfect, the secured funding has gone some way toward rebalancing the resource distribution in favour of the most marginalized groups in the process, the NGOs. At this time, the ultimate effects of this resource reallocation on nuclear fuel waste management processes in Sweden remain to be seen. As Elam and Sundqvist (2007) note, to date the Waste Network is the only NGO that is a "key player." A review of their report suggests that outside the Swedish nuclear establishment the predominant focus has been, and continues to be, on "local" municipalities – that is, geographic rather than interest-based communities.

Conclusion and Recommendations

As the Swedish example demonstrates, despite generalized theories about subpolitics and public participation, as well as more specific concepts such as the "Oskarshamn model," effective consultation for all interested communities is not guaranteed. In the Swedish situation, effective consultation has been much easier to actualize for geographic communities than for interest-based communities. This situation contrasts with the Canadian case, wherein the lack of site-specific investigations has made it extremely difficult for "local" communities to participate. Instead, environmental NGOs have provided the critical voices from the margins. It is also important to keep in mind that Swedish municipalities, when compared with Canadian municipalities, are somewhat more favourably positioned legally, politically, and financially to make independent decisions (although the interest-based community's criticisms must not be discounted). As the Canadian process moves forward and siting takes place, it will be important to avoid the

Swedish model of marginalizing the NGOs and to be mindful of the limited power and resources of Canadian municipalities as well as the limited resources of Canada's Aboriginal peoples. A more inclusive approach would be to develop processes and provide resources to all communities that wish to participate.

Other aspects of the problem for any locally affected community, such as Oskarshamn, or any geographic community that volunteers to host a nuclear fuel waste facility, are that (1) there is a tension between providing a real voice in decision making while facilitating broader policy debate and both interest-based and geographic community involvement in the development of national waste management strategies and that (2) national policies regarding the future production of nuclear waste affect the dimensions and certainty of the externality facing the local community. For instance, in the Swedish case, the nationally determined phase-out policy provides all communities with at least some understanding regarding how much waste will need to be managed, while in Canada it is clear that the ongoing commitment to nuclear power means that the amount of waste to be managed has yet to be determined. In terms of the tension outlined above, the Oskarshamn principles insist on broad participation both in national forums and in local decision making. Although this approach also seems to be sensible for local Canadian communities, particularly municipalities, it leaves us with a conundrum. How do we engage municipalities and other communities with local interests in national-level reviews and conceptual studies of nuclear fuel waste issues when their stakes in such strategies are at best unclear? Yet, if these local players are not provided with a voice, the policy that is developed will not necessarily reflect their needs and preferences. This dilemma is at the core of ideas about consultative processes and the fair distribution of risks and externalities across the landscape and society. All types of communities – at all scales – should be able to participate in decision-making processes that affect them.

In terms of justice issues, other concerns are related to the sociospatial relationships among various groups. Blowers (1999) insists that many municipalities that support the ongoing operation or the development of nuclear power and waste facilities are peripheral areas that have few other options. As outlined in this chapter, this situation is evident in both Canada and Sweden. Oskarshamn, Kincardine, Deep River, Blind River, et cetera are all dependent on the financial stability provided by the nuclear industry. Is it any wonder that these municipalities are often willing to host yet more nuclear facilities or that other peripheral areas may want to volunteer? This situation suggests that dissenting voices from both within and outside the community, such as local protest groups, national NGOs, or Aboriginal organizations, are an important avenue of deeper reflection on the costs and benefits associated with a potentially risky undertaking. At minimum, what can be done for *all* of these communities is to provide them with the resources that they need to evaluate adequately any proposed strategies that may affect them. This will involve not only economic funding but also full access to information, personnel support, a meaningful role in the decision-making process, and any other support needed to facilitate the participation of all interested communities. Although the NWMO's (2005c) siting principles support these ideas, at least at the conceptual level (see Chapter 9, this volume), it will be important to carefully evaluate the extent to which the rhetoric translates into tangible, equitable outcomes for the various communities eventually involved in any siting processes.

As this chapter illustrates, regardless of the strategy ultimately selected, it is clear that there will be impacts on existing community relationships. They may take the form of strain and hard feelings among communities or the development of new protest groups. As Canadian nuclear fuel waste management moves into a phase of emplaced decision making, it will be important to develop an in-depth understanding of the underlying nature of the communities involved and their inter-relationships, to monitor the effects of new undertakings, and to mitigate, to the extent possible, any negative community impacts.

It is also important to highlight the impact of the fluid and overlapping boundaries among and between communities on waste management processes. In the Canadian context, beyond the ambiguous place of NGOs and other national-level communities in nuclear fuel waste management processes, the overlapping jurisdictions between municipal boundaries and various Aboriginal territories (e.g., both Métis and First Nations may have claims in the same region) also require scrutiny. Historically, this overlap has occurred during the siting of numerous facilities associated with the nuclear chain, including mining, ore processing, and electricity/nuclear fuel waste production. In the future, if a facility is proposed for an area wherein municipal and Aboriginal interests and borders overlap, it will be imperative to balance more equitably these perspectives and to develop an inclusive consultative process that acknowledges the public interest as well as both municipal and Aboriginal governance structures.

Finally, given the insights gained from the Swedish case study and the various nuclear power and waste management processes completed to date in Canada, it is possible to broadly delineate a model of the various geographic and interest-based communities that will potentially become involved and influence future nuclear fuel waste management undertakings in Canada. The model (outlined below) is divided along a continuum from those with the most obvious influence and power (the nuclear establishment) through to those with the least.

Despite the imposition of a nuclear fuel waste management strategy considered to be logical and rational by the nuclear establishment, the wild cards in this model are the locally based geographic and interest-based communities that may or may not delay or derail the process once siting begins (see, e.g., Blowers 1999; and Gerrard 1994). This is implicitly acknowledged in the NWMO's (2005c, 227) final report, where the NWMO carefully delineates the equity and fairness that will be built into the siting process (see Chapter 9 in this volume) and specifically outlines the important local communities that will be consulted once siting begins. Interestingly, the NWMO focuses on local geographic communities and gives scant attention to interest-based communities either at local or at broader scales; the latent power of these local players is clearly acknowledged.

On the one hand, the legacy of past nuclear decisions, combined with the strident opposition of many municipalities (e.g., Timmins) and NGOs (e.g., Northwatch), suggests that significant local opposition will be mounted once site selection processes begin. On the other hand, to the extent to which local spaces and players are economically deprived or financially tied to the nuclear industry (e.g., Blind River), support for a nuclear fuel waste facility may also be forthcoming (see Murphy and Kuhn 2006). It is likely that some aspects of both of these positions will emerge, causing tensions within local spaces and leading to potentially significant impacts on the realization of any nuclear fuel waste management policies.

Model of Geographic and Interest-Based Communities

Nuclear Establishment (Players with Central Influence)

Central Players
- NWMO
- Natural Resources Canada (NRCan), other federal agencies, and the federal cabinet
- Regulators and others that provide oversight, including CNSC, Transport Canada, and Environment Canada
- AECL
- Nuclear power generators (Ontario Power Generation [OPG], Bruce Power, New Brunswick Power [NBP], and Hydro-Québec)

Secondary Players
- Provincial governments and their agencies
- International organizations such as the International Atomic Energy Agency (IAEA) and Nuclear Energy Agency (NEA)
- Firms that provide goods and services to the nuclear industry or in some other way have an interest in the issue
- Unions representing industry workers

Affected Actors: Governments and Civil Society Communities (Players with Variable Levels of Influence)

Central Players
- Aboriginal governments/organizations and their representatives, including the Assembly of First Nations (AFN), Métis Nation of Canada, Congress of Aboriginal Peoples (CAP), et cetera
- Directly affected regional and local government authorities and agencies as well as Aboriginal band councils and Métis communities (e.g., municipalities/Aboriginal areas with existing nuclear facilities, host, adjacent, transportation, downwind/downstream municipalities)

The Public Divide
- Directly affected geographic communities (e.g., neighbourhood closest to proposed facility, family with trapline in proposed siting area)
- Directly affected interest-based communities located within the affected region (e.g., Aboriginal women's group, business communities, local chapters/organizations oriented toward service, environmental, women's, religious, or justice activities, etc.)
- Communities of interest with broader concerns about nuclear power and waste issues (e.g., regional/provincial/federal/ international-level NGOs; Aboriginal, business, service, women's, religious, or justice groups; academics, professional organizations, experts in social or technical issues, etc.)

Players with Little to No Influence (the Silent Majority?)
- Geographic communities not directly affected by the project but nevertheless wanting to participate
- Members of the general public or other communities who may or may not be organized but only want basic information, not involvement
- Communities that do not know their interests are being affected
- Communities that do not wish to get involved at any level
- Future generations

Notes

1 There are many other ways to distinguish between communities. See, for instance, Al-Haydari (2007); and Anderson (1983).
2 The opposite scenario could also occur. In Carlsbad, New Mexico, where the majority of the community supports the military nuclear waste disposal facility, those who opposed the facility were ostracized (Murphy and Kuhn 2006).
3 For an excellent review of the Swedish case, see Sundqvist (2002).
4 It is important to keep in mind that the "Oskarshamn model" is not coterminus with a "Swedish model." Each Swedish municipality that has been involved in SKB's various studies has done so in its own way.
5 For a listing of some of Sweden's nuclear protest groups, see http://www.radwaste.org/ngo. htm.
6 For a similar discussion, see Blowers (1999).

9

Situating Canada's Approaches to Siting a Nuclear Fuel Waste Management Facility

Brenda L. Murphy and Richard Kuhn

This chapter describes the Nuclear Waste Management Organization's (NWMO) preferred management option and approach to siting, situates its proposed strategies within the international context, and assesses the challenges and opportunities presented by these strategies. Following an intensive, three-year public consultation program (see Johnson, this volume), the NWMO released its final report in November 2005. In that report, it summarized its work, outlined its preferred management approach for the long-term management of Canada's nuclear fuel waste, and delineated the broad parameters of its future undertakings. The federal government accepted the outlined approaches and preferred option, Adaptive Phased Management (APM), in June 2007.

There is no doubt that the NWMO's strategies appear to reflect state-of-the-art practices and theories, developed both here and abroad, including its focus on deep geological disposal, for the siting of a nuclear waste facility and the volunteer siting method. Nevertheless, we suggest that the nuances of the program that has been developed for the Canadian context have a range of implications for equity and short- and long-term decision making. We also argue that, despite a best practices approach, the siting process will be fraught with conflict and that the successful siting of a facility will remain far from guaranteed.

We outline and assess four aspects of the NWMO's proposed approach to the long-term management of nuclear fuel waste. First, we assess the accepted approach, APM; second, we explore the issues associated with failing to choose an economic region; third, we assess the NWMO's addition of Ordovician sedimentary rock as a potential medium for a deep geological facility; and fourth, we analyze the NWMO's chosen approach to siting (voluntary siting), wherein a community "volunteers" to host a facility. Throughout the chapter, we situate the NWMO's approach within the Canadian geographic and historical contexts as well as within the international milieu in order to tease out the implications of the NWMO's proposed plans for nuclear fuel waste management and facility siting. Among other things, the chapter focuses on the difficulty of decision making over a 325-year time horizon, the lack of place-bound representation resulting from the failure to identify economic regions, the equity issues associated with including Ordovician sedimentary rock as a geological medium, and the decision-making implications of utilizing the volunteer siting approach.

Adaptive Phased Management
As pointed out in Chapter 6 (NWMO 2005c), the NWMO chose to amalgamate the ideas of centralized storage and geological disposal by developing a fourth management option: APM. This option is essentially centralized storage followed by geological disposal over a protracted time frame.

Geological disposal is recognized internationally, at least among the nuclear establishment (see NWMO 2005c; Chapter 8, this volume), as one of the most viable management approaches,[1] with Finland and the United States having already chosen sites. Sweden is in the process of final site characterization, and several other countries, such as Switzerland and Russia, are committed to the deep geological concept (McCombie and Tveiten 2004). Furthermore, retrievability and monitoring are also prominent features of these various international processes. Although many other countries, such as Sweden, Switzerland, and Germany, have centralized storage facilities, this is not universal, and the NWMO acknowledges, quite rightly, that this element will only be incorporated if Canadians so desire. Thus, in general, the NWMO's long-term management plan (keeping in mind the different time frames attached to each country's plan) tends to dovetail closely with those of other nations; it certainly represents the predominant perspectives of the nuclear establishment (Jacob 1990; see also Chapter 8, this volume). As we outline below, it also continues Canada's historical commitment to the geological disposal option. What is distinctive about the NWMO's approach is the 325-year time horizon – our review of available sources suggests that this appears to be longer than the time frames of other countries considering deep geological disposal. The Netherlands has investigated long-term, above-ground storage (not disposal) for the next 300 years, and some countries have adopted a 100- to 150-year time horizon for disposal (McCombie and Tveiten 2004; Nuclear Energy Agency [NEA] 2001). Even if only the opening date for the deep geological disposal facility is considered, the NWMO estimates that it would not occur for seventy years (2005c, 316). This is further into the future than estimates for other countries, including the NWMO's own assessment of international approaches to deep geological disposal (2004a, 53).

From a technical perspective, the NWMO's criteria for deep geological disposal are essentially the same as Atomic Energy of Canada Limited's (AECL) concept for deep geological disposal. The NWMO's criteria include

- location in suitable rock such as the crystalline rock of the Canadian Shield or the Ordovician sedimentary rock basins;
- absence of known potential economic resources at depth;
- sufficient surface area for receipt facilities and associated infrastructure;
- seismically stable region with low known or projected frequency and magnitude of earthquakes;
- low frequency of major groundwater-conducting fracture zones, features, or faults at repository depth;
- geotechnically suitable host rock formation below the surface for the optional shallow rock cavern vaults;
- geotechnically suitable host rock formation at least 200 metres below the surface with a preference for a suitable host rock formation between 500 and 1,000 metres below the surface for the underground characterization facility and the deep geological repository;
- geochemically suitable (e.g., reducing) conditions in groundwater at repository depth;
- evidence of rock mass homogeneity and stability at repository depth;
- low hydraulic gradient and low permeability; and
- diffusion-controlled transportation of dissolved minerals at repository depth. (2005c, 233)

Thus, despite the NWMO's assertions to the contrary (see 2005c, 33), its technical approach to deep geological disposal remains essentially unchanged from the earlier AECL model, although it does incorporate the potential for centralized storage, extends the time frame, and integrates retrievability and monitoring. Canada's nuclear fuel waste management program has been committed to geological disposal since the Hare report in 1977 (Aiken, Harrison, and Hare 1977). Those outside the nuclear establishment, namely various northern municipalities, environmental non-government organizations (NGOs), and Aboriginal groups, have long disagreed that this is a viable option (see Canadian Environmental Assessment Agency [CEAA] 1998). Although the NWMO's extensive public consultation program has influenced the nuances of the long-established approach by including the features mentioned above, it has not changed it in any fundamental way (see also Johnson, this volume). Just as Sundqvist (2002) outlined for the Swedish case, although Canadian nuclear fuel waste management policies have shifted to incorporate (some would argue co-opt) a wider range of views, the fundamental locus of the policies and decision making remains firmly entrenched within the nuclear establishment (and see Durant 2007a).

Moreover, the NWMO's protracted siting time frame allows government officials to claim that they have a concrete management plan for the nuclear fuel waste "problem," even though the facility would not begin to accept waste until well into the next century. Although we agree that a well-designed, flexible adaptive process will be time consuming, the suggested 325-year time horizon essentially diffuses the potentially strident political ramifications of developing nuclear fuel waste siting policies until well beyond the terms of office of current decision makers.[2] Nuclear fuel waste management officials can, consequently, be seen to be making policy decisions without having to deal with the negative fallout often associated with decisions of this type. Critics further argue that, despite the long time horizon, the plan will allow nuclear supporters to claim that they have solved the waste problem and, hence, enable the continued expansion of the nuclear industry (Murphy 2001). Indeed, Natural Resources Canada's (NRCan) news release announcing the adoption of the APM approach states that "nuclear power is a clean energy source that emits virtually no greenhouse gases. It contributes 15 percent of Canada's electricity generation and 50 percent of the electricity supply in Ontario alone. *Nuclear energy is important to Canada's energy supply, to our security and to the Government's commitment to clean energy and the reduction of greenhouse gases*" (2007; emphasis added).

Canadian anti-nuclear groups have long argued that nuclear fuel waste management decisions must be made within the context of an integrated energy policy, in which broad debate and consultation occur regarding what Canada's energy future *should* look like. Unfortunately, despite years of protest, a promised debate, and pressure exerted by the Seaborn panel (see Durant, this volume), this issue has remained unaddressed, and the views of anti-nuclear groups remain marginalized (CEAA 1998; Murphy and Kuhn 2001).

We also argue that the long time horizon seems to be rather arrogant; it requires the designers of nuclear fuel waste policy to project three centuries into the future and impose current decisions and approaches onto future generations. Looking back from today, this would be the equivalent of the early North American French explorers or the Iroquois Confederacy developing policies for 2009! This approach appears to shift the burden of dealing with the nuclear waste problem onto future generations. If we are correct, then this contravenes the long-standing nuclear industry position that today's generation should deal with the waste problem to avoid issues of intergenerational equity (see, e.g., AECL 1994a; and NWMO 2005c).

Economic Regions

Another important aspect of the NWMO's strategies involves issues pertaining to the idea of "economic regions." The Nuclear Fuel Waste Act (NFWA) requires the NWMO to outline, in its final report, the preferred management option and to specify the economic region within which each option would be sited. The legislation states that

(3) The study must include a detailed technical description of each proposed approach and *must specify an economic region* for its implementation.

(4) Each proposed approach must include a comparison of the benefits,
risks and costs of that approach with those of the other approaches, *taking
into account the economic region* in which that approach would be implemented, as well as ethical, social and economic considerations associated with
that approach. (Government of Canada 2002, 12.3, 12.4; emphasis added)

There are seventy-six economic regions in Canada. An economic region
is a grouping of complete census divisions created as a standard geographic
unit for analysis of regional economic activity. Within Quebec, economic
regions are designated by law. In all other provinces, economic regions are
created by agreement between Statistics Canada and the provinces concerned
(Kuhn and Murphy 2003, 5).

Internationally, it appears that Canada is the only country choosing to
adopt this idea of predetermining a region using economic criteria. Other
countries have used technical criteria (e.g., the Sellafield process in Great
Britain in the 1980s), political expediency (e.g., the unilateral designation
in the United States of Yucca Mountain, Nevada, as the only site to investigate for a geological facility), or calls for potential host communities (e.g.,
Sweden and Finland) to identify potential geographic areas suitable for a
facility (McCombie and Tveiten 2004). The NFWA appears to require a mixed
approach, where an economic region is identified by the proponent, the
NWMO, and then a volunteer host community is sought within that region.
This is closer to the Swiss approach, wherein the most technically feasible
region was first identified and then a willing host was identified within that
region (Nagra 2006). The difference is that the Swiss approach used technical
criteria associated with the geological characteristics of the country to identify the most suitable region (this is an example of the top-down siting
approach described below), whereas the Canadian approach seems to be
predicated on social criteria, such as economic and ethical considerations.
Given the uniqueness of the Canadian plan and its implications for the
purported commitment to the volunteer siting method (see below), we have
questioned this approach since the draft legislation was first tabled (see
Kuhn and Murphy 2006; and Murphy 2003), and we continue to have some
concerns.

First, despite the imperative specified in the legislation, in the NWMO's
final report no economic region is specified; in essence, the NWMO appears
to have attempted to circumvent this aspect of the NFWA. The NWMO
maintains that the NFWA *does not* require it to identify a region and that it
is appropriate *not* to select a region because (1) storage on site would require
implementation in a number of regions; (2) centralized approaches, due to

transportation, require implementation in more than one region; (3) since actual site characteristics are important, it is difficult to propose an economic region prior to site investigation; (4) screening out potential sites by prematurely identifying economic regions may eliminate such sites; and (5) "narrowing the number of economic regions at this time may unduly remove communities that might otherwise wish to be considered as potential host locations" (NWMO 2005c, 145). As outlined in our 2003 background report for the NWMO on economic regions and in earlier consultations about the proposed legislation, we have always maintained that the use of this concept in the legislation and the premature designation of economic regions are problematic for the reasons outlined by the NWMO (Kuhn and Murphy 2003). Yet the legislation clearly states that the NWMO must designate such a region. Thus, although we agree with many of the NWMO's reasons for not wanting to select an economic region, we maintain that its approach contravenes the legislation by not identifying these regions. This not only continues the nuclear industry's policy of placelessness (see Murphy, Chapter 8, this volume) but also means that the NWMO, as required by the legislation, was unable to identify the benefits and costs of the various approaches relative to the identified economic region.

Second, the failure to identify economic regions denies geographic communities a voice on the Advisory Council. This council was mandated in the NFWA to provide the NWMO with feedback from an independent panel of experts representing a wide range of interests, including the affected economic region. According to the legislation, the council should include "representatives nominated by local and regional governments and aboriginal organizations that are affected because their economic region is specified for the approach that the Governor in Council selects [e.g., the preferred management option]" (Government of Canada 2002, 8.2.c). Failing to select an economic region denies this required local/regional representation and marginalizes these voices.

Ordovician Sedimentary Rock

The NWMO's final report states that the proposed facility will be sited in either the Canadian Shield or Ordovician sedimentary rock. Given that Ontario, New Brunswick, Quebec, and Saskatchewan are the provinces directly involved in the nuclear chain (including uranium mining, ore processing, and nuclear electricity generation [NWMO 2005c]), the proposed facility will probably be located within the boundaries of one of these provinces. We suggest that it is uncontroversial and equitable that facility siting should occur within those provinces that have most benefited from the employment and electricity generated by nuclear activities. Far more controversial is the NWMO's assertion that a facility could be sited in Ordovician sedimentary rock. As the Advisory Council states in its review of the NWMO's study, "the option to use sedimentary rock was introduced relatively late in NWMO's study process, and limited work has been undertaken in Canada to date on Ordovician sedimentary rock to determine its suitability for this purpose. It is therefore premature to consider Canadian sedimentary and crystalline [granitic] rock as equivalent options until more research has been undertaken on the former" (NWMO 2005c, 446).

Until this final report, Canadian nuclear fuel waste policy was firmly focused only on the granitic rock of the Canadian Shield. Granitic rock was considered the most viable Canadian geological medium due, among other things, to its stable, relatively unfractured, nature (AECL 1994a).[3] Sedimentary rock has never been seriously researched in Canada, although the NWMO (2006) suggests that this is currently being assessed in France and Switzerland. Other geological mediums also being considered worldwide are clay (Switzerland, France), salt domes (Germany), and volcanic tuff (United States). Some countries are considering more than one medium. For instance, Germany is investigating salt, granite, and clay (NEA 2001). Canada, like most countries with access to granitic rock, such as Sweden and Finland, considers it to be one of the most important rock formations to investigate. In its final report, the NWMO notes that, despite some initial research suggesting that Ordovician rock may be a suitable geological medium, "more research and development work on sedimentary rock needs to be completed to determine the suitability of these formations" (2005c, 135). Then, in its 2006 annual report, the NWMO asserts that "technical research [in 2006] focused on assessing the current state of knowledge on the technical feasibility and safety of sedimentary rock as a potential host rock formation for a deep geological repository" (22). It seems odd that sedimentary rock formations would be included as an option on par with granitic rock formations in its 2005 final recommendations to the government when the feasibility of the Ordovician option has not yet been established.

Two key reasons, related to equity issues, can be suggested for the inclusion of Ordovician sedimentary rock. First, since the Canadian Shield is located only in the northern parts of Canada, the inclusion of sedimentary rock as a potential geological medium means that both southern and northern regions can now be considered. On the one hand, this is more inclusive and equitable in that all economic regions, particularly those within the designated provinces, can now be included among potential host sites. On the other, siting of what is often perceived as southern Canada's waste in northern areas has been a point of contention, an issue of environmental justice, among NGOs, Aboriginal nations and organizations, and northern municipalities (Murphy 2001). Including the geology of southern Ontario serves to dissipate the force of this critique and derails one of the key arguments often raised by opposition groups.

Second, many of the communities that currently host nuclear facilities in Ontario – such as the nuclear power generating stations at Pickering, Darlington, and Kincardine – are located in southern Ontario, areas of sedimentary rock. Since there is evidence that communities already hosting nuclear facilities may be predisposed to hosting additional facilities (see, e.g., Blowers 1999; and Sundqvist 2002),[4] the inclusion of sedimentary rock areas strategically opens up this possibility for Canada. There is already some evidence of this in Ontario, where the municipality of Kincardine, located adjacent to the Bruce Power nuclear energy generating plant, has agreed to support the investigation of a proposed low-intermediate nuclear waste facility within its boundaries (Al-Haydari 2007). Given that the viability of the Ordovician sedimentary rock option is, at this time, speculative at best, we question to what extent the option was included for political expediency both to dissipate arguments and increase the possibility that a community will step forward to host the proposed facility.

The NWMO's Siting Approach

In this section, we provide the background within which to understand the NWMO's approach to siting, which is focused on volunteer host communities. Essentially, its approach follows from that developed by AECL (1994a) and the Seaborn panel's report (CEAA 1998). Since the NWMO has chosen to ignore the economic region designation, its approach also closely aligns with what is often thought of as "best practice" for the siting of a wide range of noxious facilities. The NWMO's approach is similar to that implemented or proposed by most countries, including Finland, Sweden, Japan, and France (McCombie and Tveiten 2004). Despite the wide-scale adoption of this approach, we maintain that it will not necessarily lead to a non-conflictual siting process or to the siting of a facility.

The siting process associated with implementing the NWMO's preferred approach can be characterized as a "bottom up," open, voluntary process in which a willing community is sought to host the proposed facility. This approach was employed in Alberta in 1987 to site a hazardous waste facility at Swan Hills. This "successful" siting has long been held up as a model for other facility siting exercises.[5] It was also used in the failed historic low-level nuclear waste facility siting process in Deep River, Ontario (Hunold 2002; Murphy 2001). Internationally, it was attempted in Sweden, but ultimately the communities with existing nuclear facilities had to be invited to participate before two potential host communities were eventually identified (Sundqvist 2002). Nevertheless, virtually all countries (except the United States) profess to adhere to some version of the voluntary approach for the siting of nuclear waste facilities (McCombie and Tveiten 2004).

In the voluntary siting approach, the proponent invites communities to learn more about a particular project initiative and then allows them to decide whether they would like to host the facility (Ballard and Kuhn 1996). Notice here that the term "communities" is used in the sense of place-based or geographic communities (see Murphy, Chapter 8, this volume); the role of interest-based communities, for instance NGOs, is not clearly accommodated in the volunteer model. Communities choose to become involved, have access to information and financial resources to help them make informed decisions, and can withdraw from the process at any time. In an ideal situation, more than one community with the needed technological characteristics to ensure safety decides that the benefits of the project outweigh the drawbacks, and these communities compete for the right to host the facility (Wolsink 1994). The voluntary model is premised on the idea that siting will result in an acceptable site rather than the best site. An acceptable site is interpreted as technologically feasible, safe for both the built and the natural landscapes, and satisfactory to the locally affected community. The open approach provides a set of safeguards to protect the rights and safety of the community. These safeguards include assurances that agreed-on minimum health, safety, and environmental standards will not be breached, that communities can negotiate mitigation and compensation measures, and that there will be full information disclosure and provision of technical and financial support for the community (Armour 1992).

This approach to siting was first developed in the late 1980s in response to ongoing siting conflicts and failures whenever attempts were made to site locally unwanted land uses (LULUs). These siting attempts were typically reported to cause public outcry and the development of opposition groups, particularly for those located closest to the facility. This opposition is said to lead to a not in my backyard (NIMBY) attitude of local residents who perceive that the facility will unfairly burden their local area (Schively 2007).

However, more detailed understandings of NIMBY attitudes have revealed that there are often other reasons for this opposition. First, these facilities may lead to an inequitable distribution of costs and benefits, with the benefits accruing to the broader society, while the costs are concentrated within the siting community (Armour 1992; Wolsink 1994). Second, the NIMBY label often assumes that there is widespread agreement on the need for and usefulness of the facility. Third, the NIMBY label denies that LULUs are also associated with lack of trust in proponents and regulators and feelings of loss of control within the affected area (Kraft and Clary 1991).

Approaches most often associated with LULUs and NIMBYs tend to be linked to a top-down siting model dubbed "decide-announce-defend" (DAD) (Kunreuther, Fitzgerald, and Aarts 1993). In this approach, the siting organization uses technical criteria to narrow the search from a broad investigation to specific site characterization. The site selection process begins when a large region is screened, using technical criteria such as geological and hydrological data and settlement patterns. The aim is to locate a number of areas that could be potential sites (Armour 1990). This is similar to the process undertaken by the Swiss to identify the region for a nuclear waste facility. At the screening stage, readily available secondary source information is used. Typically, top-down siting then narrows to a more in-depth investigation, within the identified areas, and the eventual identification of the "best" site. However, defending the "best" site is virtually impossible since there will always be debate regarding the selection criteria and the process by which the site was selected. Debates surrounding definitions of the "best site" lead invariably to the local questioning of technical expertise. Furthermore, in this approach, there is little consultation with the public, and decision making lies with the proponent and oversight agencies. The lack of substantive public involvement engenders distrust in the proponent and loss of legitimacy (Armour 1992).

In Canada, an example of this criteria-driven locational approach, focusing on the goal of finding the best possible site, occurred in Ontario during the Ontario Waste Management Corporation's (OWMC) search for a hazardous waste facility. The painstaking process took four years and $10 million to complete (Armour 1990). Yet, in the end, the siting attempt was a failure as the Ontario Environmental Assessment Board did not accept the assessment. The top-down, DAD approach was also used by the Nuclear Industry Regulatory Executive in the United Kingdom to locate potential disposal sites for nuclear fuel waste (Openshaw, Carver, and Fernie 1989). Tellingly, this process eventually had to be abandoned due to public outcry. Freudenburg (2004) maintains that, for the siting of nuclear facilities in the United States, the top-down approach generally led to failure.

Thus, the NWMO's approach, based more or less on the volunteer siting model, attempts to address many of the problems identified with top-down initiatives and LULUs and reflects up-to-date thinking regarding the involvement of communities in the siting process. In developing its approach to implementing the preferred option, the NWMO states that it will involve people in a collaborative exercise to design and implement the public engagement process. Its proposed strategy involves the following:

- It seeks input from the potentially affected communities of interest concerning the way in which they wish to be engaged in the process. It initiates an open dialogue with affected interests, both to invite comment on progress to date and to help shape future plans.
- It seeks direction and input from potentially affected communities regarding how to analyze possible socioeconomic effects of the implementation activities on a host community's way of life or on its social, cultural, or economic aspirations and considers how best to manage those effects.
- Working with potentially impacted communities, it undertakes research into social and ethical considerations and impacts that will be encountered

through implementation and operation of the management facility. It undertakes research on adaptive management as it relates to ongoing social and technical decision making, including research to support the identification and management of possible community impacts, such as impacts on traditional Aboriginal lands.

- It supports the development of capacity in potentially affected communities so that they may increase their understanding of implementation issues and initiate their own investigations as required. (NWMO 2005c, 234)

In terms of siting, in its final report, the NWMO states the following:

1 It will seek an informed, willing community to host the central facilities. The site must meet the scientific and technical criteria chosen to ensure that multiple engineered and natural barriers will protect human beings, other life forms, and the biosphere. Implementation of the approach will respect the social, cultural, and economic aspirations of the affected communities.
2 It will sustain the engagement of people and communities throughout the phased process of decision and implementation.
3 It will be responsive to advances in technology, natural and social science research, Aboriginal traditional knowledge, and societal values and expectations. (2005c, 44)

According to the NWMO (2005c, 74), the ethical values associated with this process include respect for life, future generations, the biosphere, and peoples; justice and fairness among groups, regions, and generations, especially for those affected; and sensitivity to value differences among participants.

The NWMO's proposed approach follows closely from that delineated by AECL, as laid out in its environmental assessment of the deep geological disposal concept. According to AECL, siting should be based on the following:

1 *Volunteerism*: The community has the right to determine its willingness to host a facility. The community must have jurisdiction over that territory. Where crown land is involved, consent from the government with jurisdiction will be sought, and that government will be encouraged to identify a potential host.
2 *Shared decision making*: Implementation would occur in stages with series of decisions. The potential hosts will share in decision making, as negotiated. The proponent will seek to address the views of other potentially affected communities.

3 *Openness*: The implementing organization would offer information about its plans, procedures, activities, and progress early in the siting stage and would continue to do so. The community would have access to all information it requires to make a judgment about safety and environmental protection.

4 *Fairness*: Since the host will be providing a service to electricity consumers, the net benefit to the host should be significant. Measures will be taken to avoid, mitigate, and compensate for adverse effects and to enhance and ensure the betterment of the community. Fairness also requires due process and adherence to the above three principles. (1994b, 23-24; emphasis added)

The Seaborn panel's report (CEAA 1998) reiterated these principles and added the following criteria: the willingness to be involved as a potential host does not represent a final commitment; monitoring and compensation proposals should be developed early, in consultation with communities, to reduce and mitigate adverse effects; the facility must meet regulatory standards; there must be adequate time provided to understand the social, technological, and environmental implications before communities make decisions; there should be early agreement regarding processes for conflict resolution; and the decision-making process should be inclusive of minorities (CEAA 1998, section 6.3.1; Murphy 2001).

The volunteer siting approach, as proposed by the NWMO and most countries around the world, was developed to allow communities to decide whether or not they wish to become involved in hosting a noxious facility. There are several purported benefits of this process. For the local community, democracy and rights may be more clearly preserved; for the proponent, it is thought that conflict may be avoided and that the siting process has a greater chance of being "successful" – that is, an actual facility will be sited (Ballard and Kuhn 1996).

However, we argue that the volunteer siting approach will not necessarily achieve these stated community or proponent goals. At the broadest scale, since the volunteer siting approach defines the management of noxious wastes, such as nuclear fuel waste, as a management "problem" looking for a "local" solution, questions regarding the continuing generation of the waste are deemed inadmissible or outside the proponent's mandate (Lake and Disch 1992). For instance, in its final report, the NWMO asserts that "we report below [in Chapter 4] on some fundamental questions on which we heard the views of Canadians diverge. *For the most part, these questions are beyond the mandate of our study.* However, the divergence of views on these questions infuses many of the comments we heard about the management approaches. The differences in perspective on these questions are important influencing factors, which the study must recognize, although it cannot directly address them" (2005c, 71; emphasis added).

One of the questions that the NWMO (2005c) considers outside its mandate is the concern regarding whether or not Canadians should continue with the generation of nuclear electricity. Yet the earlier Seaborn panel's report (CEAA 1998) clearly included some aspects of this question despite a similarly narrow mandate (Murphy and Kuhn 2001). This was also an issue raised in the Advisory Council report to the NWMO: "The potential role of nuclear energy in addressing Canada's future electricity requirements needs to be placed within a much larger policy framework that examines the costs, benefits and hazards of all available forms of electrical energy supply, and that framework needs to make provision for comprehensive, informed public participation" (NWMO 2005c, 436).

Furthermore, as we have already indicated, decisions about facility siting can have national energy policy implications. We maintain that the ongoing tension around the marginalization of this issue could potentially delay or derail any attempted siting process based on the narrowly defined volunteer siting process (see also Kunreuther, Fitzgerald, and Aarts 1993). As Murphy and Stanley (2006, 5) assert, "these broader social questions cannot be answered during a volunteer siting process that defines the waste problem as being about finding a safe and acceptable site. And yet, during many siting processes, these larger issues interfere with the capacity of the project implementers to actually site their desired facility or achieve their policy goals. One of the inherent problems is that the agencies mandated with managing the waste work within a pre-determined framework (usually focused on a siting solution) and feel unable or unwilling to push the limits of those boundaries."

Moreover, beyond these broader issues, the details of equitably implementing the voluntary siting approach are also complex and politically fractious, potentially leading to siting conflict or failure. For instance, during the search to find a host for Ontario's historic low-level nuclear waste, the local "community" was originally defined to include Deep River and several of the neighbouring municipalities. However, as the process progressed, successive municipalities opted out, and eventually only Deep River remained as the potential host community. This occurred even though the proposed facility was located quite distant from the municipality's populated area but very close to the main village in the adjacent community of Chalk River (Hunold 2002).

The following is a summary of some of the other potential problems and issues that need to be addressed within a siting process utilizing the volunteer siting method (see Murphy 2001). As will be immediately obvious, there are myriad opportunities for conflict and siting failure as the context-specific details of the volunteer approach are negotiated.

1 The volunteer siting process often assumes that the municipality within which a facility will be sited should be considered the host community. But this leaves a plethora of questions unanswered. How will the definitions of, and levels of impact on, the host, affected (adjacent), downstream/downwind, and transportation communities be defined? Who will define these communities and set their boundaries? What criteria will be used to set these boundaries (e.g., political, economic, social, and ecological criteria)? Who will represent them? How will their various interests be balanced? How will the differing boundaries and interests of Aboriginal and non-Aboriginal peoples, who may be present in some areas, be balanced and resolved? How will the views of different interests within the communities be accessed and integrated into the dialogue? How will conflicts among different interests, both within and between communities, be resolved?

2 To what extent can the facility be engineered to overcome the specific technical and safety-related characteristics of the volunteered site? Sundqvist (2002) argues that in Sweden, since it proved so difficult to get any community to consider hosting a facility (despite a national nuclear phase-out policy), attention has shifted toward engineering around any geological difficulties that may present themselves at the proposed host's site rather than adhering to the preset technical criteria. What does this imply about the inherent safety of the facility?

3 How can a balance be achieved between the right of a community to have self-determination and the concern that some communities may volunteer due to economic blackmail? How can fair compensation levels be determined and implemented without appearing to be a form of bribery? For instance, in the case of Deep River, the main "selling feature" of the proposed facility was the guarantee of job maintenance at the nearby Chalk River Laboratory (Hunold 2002; Murphy 2001). Keith Lewis from Serpent River First Nation commented on these problems in relation to siting a facility on First Nations land:

> When it comes to ... the high level waste disposal concept, the possibility of siting becomes more of a reality in the near future, and a reality that would amount to [nuclear wastes] being located on Indian lands, treaty lands. When you have those realities that we have in this community, and when you think that all Indian communities are like that, when a process like the high level radioactive waste process throws money at a community that's starving, what are they going to say? Then when you get back to that thing about morality, and there is nothing moral about buying out somebody that's starving. Rather than taking care of the social and cultural dysfunction in the community, righting that first, they just say here's a whole pile of money, you do with it what you want, and let us destroy the

environment more, and create further potential harm for your people and the land. (Reckmans, Lewis, and Dwyer 2003, 20)

4 How will community acceptance be measured? Methods currently used include surveys, elected official votes, and referendums. What will be the level of acceptance required (e.g., 50 percent + 1)?
5 In northern areas, with unorganized townships and no local government structures, who would be the host community, and who would represent it?
6 How will a "successful" siting process be defined? Defining success as the achievement of a fair, equitable process, regardless of the ultimate outcome, is quite different from a definition based primarily on the end point of finding a willing host.
7 The waste will need to be transported over long distances, often through small communities, along busy roads, during bad weather, et cetera. The NWMO (2005c, 25) estimates that movement of the projected 3.6 million fuel bundles will require fifty-three road shipments per month for thirty years. That is a total of 19,080 shipments! Transportation will inevitably be delayed by protests, bad weather, and other circumstances, thereby extending the number of years over which the movement of the waste to a central facility will affect nearby communities.[6] How will siting incorporate the needs and concerns of the people affected by these transportation routes?
8 What will be considered adequate levels of intervenor funding? Who will have access to these funds? Who gets to make these decisions? According to an AECL study, the current funding programs, provided by the environmental assessment process and Indian and Northern Affairs Canada, are clearly inadequate for facilitating Aboriginal participation in environmental reviews of proposals that affect their interests (HBT AGRA Limited 1993).
9 How appropriate is the volunteer siting approach for Aboriginal communities? According to the Seaborn panel's report (CEAA 1998), the Aboriginal community told the panel that (1) neither it nor the proponent consulted with Aboriginal people in an appropriate manner, (2) Aboriginal people had not been given enough time and their own approaches and language to study and understand the proposal, and (3) a siting process based on volunteerism does not fit their traditions and views of community, and they have a suspicion that promises will be broken.
10 How will the siting process be incorporated into the government-mandated environmental assessment processes and the Canadian Nuclear Safety Commission (CNSC) licensing processes? Which government authority will review the proposals and have ultimate decision-making

power? How will various public voices, including that of the "potential host," be incorporated and influence the process?

11 What will be the alternative if the voluntary approach cannot identify a willing host? Would the communities currently hosting the nuclear fuel waste storage facilities (e.g., Darlington, Kincardine, and Pickering) become the long-term hosts by default?

12 Does the Canadian nuclear industry have a clear understanding regarding how much waste, and in what form, a host facility would be expected to manage? Refurbishments of existing nuclear power plants are under way, and there are currently Ontario commitments to build more generating stations that may not use Canadian Deuterium Uranium (CANDU) technology. In this situation, volunteer communities face considerable uncertainty when thinking about hosting a waste facility.

Although it is clear from its final report that the NWMO is cognizant of many of these concerns, it maintains a very positive tone throughout its description of the volunteer siting approach.

> Initiatives must be designed to seek positive contributions to the community that will continue over the long term. Further, the issue is not simply one of jobs, income, or tax revenues. More fundamentally, it is an issue of people's future and the degree of confidence that this future will unfold in a manner consistent with closely held values and priorities. This touches the heart of a community's culture. If synchronicity between a proposed project and people's values is not evident, the project may be seen as a threat to the fabric of community life, and be vehemently opposed ... *We believe that such an alignment is possible.* (2005c, 277; emphasis added)

This might be the NWMO's belief, but given past experiences both within nuclear fuel waste management and in other contexts, we suspect that significant problems may develop as the NWMO begins to undertake the siting process. Furthermore, our confidence in the nuclear industry's capacity to truly undertake a co-operative, inclusive siting approach is undermined by various ongoing and past processes. The most recent process is the proposed siting of the low-intermediate nuclear waste facility in Kincardine, which, despite a willing host and access to knowledge regarding "best practices," has been fraught with controversy and animosity. Of note are the ongoing tensions regarding levels of compensation between host and adjacent communities, lack of compensation provided for the Aboriginal peoples (i.e., Saugeen and Nawash) on whose territory the facility will be located, disagreement regarding the validity of the phone poll undertaken to gauge public acceptance, and lack of any discussion regarding the continued production of this type of waste (Al-Haydari 2007). As Ontario Power Generation (OPG) is the proponent of this project, and one of the principal players on the NWMO's Board of Directors, we have grave concerns regarding the extent to which the spirit and practices of the volunteer siting approach will be utilized and whether any facility will eventually be sited.

Conclusion

This chapter has outlined four key aspects of the NWMO's proposed approach to managing Canada's nuclear fuel waste. This approach mirrors closely that supported by other nations, particularly from predominant government and nuclear industry perspectives. Although environmental NGOs and Aboriginal peoples have not, for the most part, supported deep geological disposal, and although they have continuously demanded a wider dialogue regarding the continued production of nuclear fuel waste, their concerns have been marginalized and written out of the mandate of the implementing authorities, including the NWMO.

We have outlined our concerns regarding the inclusion of sedimentary rock formations and the protracted siting time horizons associated with the adaptive phased approach. The idea of siting a facility in Ordovician rock is not currently supported by research assessing the technical viability of this option in a Canadian context. The 325-year time horizon seems to be out of step with that proposed by other countries. We wonder to what extent these two aspects of the NWMO's plans are included more for their political expediency than for their justified technical requirements.

The NWMO's refusal to designate economic regions, as required by the NFWA, denies geographic communities the opportunity to have a voice on the Advisory Council regarding the development of approaches to siting and other future nuclear fuel waste management plans developed by the NWMO. In contrast, if implemented according to the legislation, the economic region approach would undermine the tenets of the volunteer siting process by predetermining specific areas within which siting should take place. More profoundly, this confusion regarding whether or not economic regions should be identified and when adds another layer of tension and uncertainty within any attempted siting processes. It is unclear whether this confusion could lead to siting conflict or failure.

Although the issue of nuclear fuel waste transport has not been evaluated in any substantial way, it may be one of the most contentious issues to be faced by the NWMO as it moves into the implementation phase. Small-scale examples from Canada, such as the movement of MOX (see note 6), large-scale protests in Germany,[7] and ongoing concerns in the United States[8] are all instructive. They suggest what could happen in Canada once a host community site is announced and municipalities along the route begin to comprehend the direct impact of the proposed facility.

Finally, although the volunteer siting framework is generally acknowledged as providing some level of justice, self-determination, and democracy for local communities, the method is not a panacea that will guarantee an equitable, conflict-free process or the eventual siting of a facility. Included among the plethora of politically volatile problems that will need to be addressed are how a host community will be defined, how concerns and voices of other communities will be included, how issues of economic blackmail and compensation as bribery will be addressed, and how the problem of incorporating Aboriginal perspectives into a Western-defined process will be solved. These and other thorny questions will have to be resolved before Canadians can hope to find a solution to the long-term problem of managing Canada's nuclear waste.

Notes

1 It is important to keep in mind that the nuclear establishment worldwide is a small, closely knit group and that those outside this group often oppose this hegemonic perspective.

2 This is a phenomenon known as NIMTOO – not in my term of office.

3 Although even these characteristics have been questioned, see reports of the Technical Advisory Committee in association with AECL's assessment of geological disposal (AECL 1994a).

4 For instance, in the United Kingdom in June 2007, the economically depressed community of Cumbria, with ties to the nuclear industry, volunteered to be considered as a host for a high-level nuclear waste facility. See the story at http://www.bbc.co.uk/cumbria/content/articles/2005/11/29/nuclearpower_feature.shtml.

5 See Bradshaw (2003), however, for some problems with this facility and its siting, including lack of consultation with First Nations, contamination of adjacent land, and failure to restrict the intake of hazardous waste strictly from Alberta sources.

6 The fear of protest associated with moving nuclear waste is demonstrated by the Government of Canada's decision to fly mixed oxide fuel (MOX) to Chalk River, Ontario, rather than risk road transport. The MOX transport was undertaken clandestinely, without any notification to municipalities in the flight path, and it contravened the government's own policy on MOX transport. See http://www.cnp.ca/issues/.

7 See UK Indymedia, "Wendland, Germany: Resisting Castor Nuclear Transport," http://www.indymedia.org.uk/en/2006/11/355633.html.

8 For instance, Halstead, Dilger, and Ballard (2004) argue that the transport route to Yucca Mountain lies within lands claimed by the Western Shoshone Nation, may result in conflicts with adjacent ranchers, and will require movement of the waste through Las Vegas (vehemently opposed by local officials).

References

Adams, T. 2000. *From Promise to Crisis: Lessons for Atlantic Canada from Ontario's Electricity Liberalization.* Halifax, Nova Scotia: Atlantic Institute for Market Studies.

–. 2006. "Nuclear Power: Still a 'Promising' Industry. Comment on OPA Supply Mix Advice Report." Prepared for Energy Probe, submission to Ontario Power Authority (OPA), 28 February. http://www.powerauthority.on.ca or http://www.energyprobe.org.

Aiken, A.M., J.M. Harrison, and F.K. Hare (chairman). 1977. *The Management of Canada's Nuclear Wastes.* Report of a study prepared under contract for the Minister of Energy, Mines, and Resources Canada, EMR Report EP 77-76. Ottawa: Energy, Mines and Resources.

Al-Haydari, D. 2007. "Community Dynamics in the Siting Process for a Low to Intermediate Level Nuclear Waste Disposal Facility in Kincardine, Ontario." Master's thesis, University of Guelph.

Anderson, B. 1983. *Imagined Communities: Reflections on the Origin and Spread of Nationalism.* London: Verso.

Arai, A.B. 2001. "Science and Culture in the Environmental State: The Case of Reactor Layups at Ontario Hydro." *Organization and Environment* 14, 4: 409-24.

Armour, A. 1990. "Socially Responsive Facility Siting." PhD diss., University of Waterloo.

–. 1992. "*The Co-Operative Process: Facility Siting the Democratic Way. Plan Canada,*" March, 29-34.

Armstrong, C. 1981. *The Politics of Federalism: Ontario's Relations with the Federal Government, 1867-1942.* Toronto: University of Toronto Press.

Armstrong, C., and H.V. Nelles. 1986. *Monopoly's Moment: The Organization and Regulation of Canadian Utilities, 1830-1930.* Philadelphia: Temple University Press.

Assembly of First Nations (AFN). 2004a. *First Nations Nuclear Fuel Waste Dialogue Working Group Meeting #2 Report.* Submitted to the NWMO, 30 November, Ottawa, AFN.

–. 2004c. *Nuclear Waste Management Regional Forum: Northern Ontario.* Wauzhushk Onigum First Nation, 23 November. Submitted to the NWMO, 17 December, Ottawa, AFN.

–. 2004b. *Nuclear Waste Management Regional Forum: Southern Ontario.* Toronto, 18 November. Submitted to the NWMO, 17 December, Ottawa, AFN.

–. 2004d. *Nuclear Waste Management Regional Forum: Western Canada.* Prince Albert, SK, 30 November. Submitted to the NWMO, 17 December, Ottawa, AFN.

–. 2005a. Annual General Assembly, Resolution No. 39/2005 on the Nuclear Waste Management Dialogue Process. Adopted 7 July, Yellowknife, AFN.

–. 2005b. *Nuclear Fuel Waste Dialogue: Phase II Regional Forums Summary Report.* Submitted to the NWMO, 31 January, Ottawa, AFN.

–. 2005c. *Nuclear Fuel Waste Dialogue: Recommendations to the Nuclear Waste Management Organization.* Submitted to the NWMO, 30 September, Ottawa, AFN.

–. 2005d. *Nuclear Fuel Waste Management Dialogue: Final Report to the Nuclear Waste Management Organization.* Submitted to the NWMO, 30 November, Ottawa, AFN.

–. 2005e. *Nuclear Waste Management Regional Forum: Quebec.* Ottawa, 26 July 2005. Submitted to the NWMO, 13 September, Ottawa, AFN.

Atomic Energy of Canada Limited (AECL). 1994a. *Environmental Impact Statement on the Concept for Disposal of Canada's Nuclear Fuel Waste.* AECL-10711, COG-93-1. Pinawa: AECL.

–. 1994b. *Summary of the Environmental Impact Statement on the Concept for Disposal of Canada's Nuclear Fuel Waste.* AECL-10721, COG-93-11. Pinawa: AECL.

–. 1995. "Research and Development Advisory Panel to the Board of Directors of AECL." In *Compendium of Public Comments on the Adequacy of the Environmental Impact Statement on the Nuclear Fuel Waste Management and Disposal Concept,* vol. 2, Nuclear Fuel Waste Disposal Concept Environmental Assessment Panel, 1-7. Publication TEC 007. Ottawa: Ministry of Supply and Services.

–. 2002. *Report of the AECL R&D Advisory Panel for 2002.* Pinawa: AECL.

Atomic Energy Control Board (AECB). 1985. *Deep Geologic Disposal of Nuclear Fuel Waste: Background Information and Regulatory Requirements Regarding the Concept Assessment Phase.* AECB Regulatory Policy Document R-71. Ottawa: AECB.

–. 1987. *Regulatory Objectives, Requirements, and Guidelines for the Disposal of Radioactive Wastes: Long-Term Aspects.* AECB Regulatory Document R-104. Ottawa: AECB.

Auditor General of Canada. 1995. *Federal Radioactive Waste Management.* Report of the Auditor General of Canada to the House of Commons. Catalogue No. FA1 1995/1-4E. Ottawa: Minister of Supply and Services Canada.

Babin, Ronald. 1985. *The Nuclear Power Game.* Trans. Ted Richmond, Foreword by Gordon Edwards. Montreal: Black Rose Books.

Ballard, K., and R.G. Kuhn. 1996. "Testing Community Empowered Siting for Canadian Nuclear Waste." In *Proceedings of the 1996 International Conference on Deep Geological Disposal of Radioactive Waste,* CNS, Lac du Bonnet, MB. *Lac du Bonnet Leader,* 10-1-10-10.

Barnaby, J. 2003. *Drawing on Aboriginal Wisdom: A Report on the Traditional Knowledge Workshop.* Saskatoon, September. Toronto: NWMO. http://www.nwmo.ca.

Bauder, H. 2006. *Labour Movement: How Migration Regulates Labour Markets.* New York: Oxford University Press.

Beal, A., S. Boulby, J. Fowler, V. Horn, and M. Smith. 1987. *Women and Nuclear Safety Project.* Submission to the Ontario Nuclear Safety Review, chaired by F.K. Hare, on behalf of Queen's University Women's Centre, 1 September.

Beck, U. 1992. *Risk Society: Towards a New Modernity.* London: Sage.

–. 1994. "The Reinvention of Politics: Towards a Theory of Reflexive Modernization." In *Reflexive Modernization: Politics, Tradition, and Aesthetics in the Modern Social Order,* ed. U. Beck, A. Giddens, and S. Lash, 1-55. Stanford: Stanford University Press.

–. 1995. *Ecological Enlightenment: Essays on the Politics of the Risk Society.* Atlantic Highlands, NJ: Humanities Press.

–. 1996. "Risk Society and the Provident State." In *Risk, Society, and Modernity: Toward a New Ecology,* ed. S. Lash, B. Szerszynski, and B. Wynne, 27-43. London: Sage.

–. 1998. *World Risk Society.* Cambridge, UK: Cambridge University Press.

Benhabib, S. 1996. *Democracy and Difference: Contesting the Boundaries of the Political.* Princeton: Princeton University Press.

Berger, T. 2005. *Comments on NWMO's Consultation Process.* Toronto: NWMO. http://www.nwmo.ca.

Blow, P. 1999. *Village of Widows* [documentary video]. Lindum Films, Peterborough.

Blowers, A. 1999. "Nuclear Waste and Landscapes of Risk." *Landscape Research* 24: 241-64.

Bothwell, R. 1984. *Eldorado: Canada's National Uranium Company.* Toronto: University of Toronto Press.

–. 1988. *Nucleus: The History of Atomic Energy of Canada Limited.* Toronto: University of Toronto Press.

Bourdieu, P. 1991. *Language and Symbolic Power.* Trans. G. Raymond and M. Adamson. Cambridge, MA: MIT Press.

Bradshaw, B. 2003. "Questioning the Credibility and Capacity of Community-Based Resource Management." *Canadian Geographer* 47: 137-50.

Brand, S. 2003. "Thinking about Time." http://www.nwmo.ca.

Bratt, D. 2006. *The Politics of CANDU Exports*. Toronto: University of Toronto Press.

Brisco, B. 1988. *High-Level Radioactive Waste in Canada: The Eleventh Hour*. Report of the Standing Committee on Environment and Forestry on the Storage and Disposal of High-Level Radioactive Waste. Second Session of the 33rd Parliament, 1986-87-88. Ottawa: Ministry of Supply and Services.

Brook, A. 1995. "Risk Assessment and Moral Assessment." In *An Environmental Ethics Perspective on Canadian Policy for Sustainable Development*, ed. Institute for Research on the Environment and Economy. Ottawa: Institute for Research on Environment and Economy.

–. 1997. "Ethics of Wastes: The Case of the Nuclear Fuel Cycle." In *Canadian Issues in Applied Environmental Ethics*, ed. A.W. Cragg and A. Wellington, 117-32. Peterborough: Broadview Press. http://http server.carleton.ca/~abrook/NUCLEAR.htm.

Brown, P.A., and C. Létourneau. 2001. "Nuclear Fuel Waste Policy in Canada." In *Canadian Nuclear Energy Policy: Changing Ideas, Institutions, and Interests*, ed. G.B. Doern, A. Dorman, and R.W. Morrison, 113-46. Toronto: University of Toronto Press.

Brown, P.G. 1994. *Restoring the Public Trust*. Boston: Beacon Press.

Bruce Power. 2006. "Bruce Power Enters Next Phase of Long Term Siting Planning." Press release, 17 August. http://www.brucepower.com.

Brunk, C. 1992. "Technological Risk and the Nuclear Issue: Abstract of Presentation, Environmental Ethics Workshop." Institute for Research on Environment and Economy, University of Ottawa, 1-2 October.

–. 1995. "Technological Risk and the Nuclear Safety Debate." In *An Environmental Ethics Perspective on Canadian Policy for Sustainable Development*, ed. Institute for Research on Environment and Economy. Ottawa: Institute for Research on Environment and Economy.

Bullard, R. 1999. "Dismantling Environmental Racism in the USA." *Local Environment* 4, 5-19.

Burnham, C., Consulting, and R.J. Readhead. 2004. *Key Points Raised during Discussions with Senior Environmental and Sustainable Development Executives*. Toronto: NWMO.

Campbell, M. 2005. "Bury Nuclear Waste Underground, Group Says." *Globe and Mail*, 4 November, A8.

Canada. House of Commons. 2001a. *Debates*, 8 May.

–. 2001b. *Debates*, 5 December. David Pratt, Nuclear Fuel Waste Act, third reading.

–. 2002a *Debates*, 26 February.

–. Senate. 2002b. *Debates*, 13 March. Lois M. Wilson, Nuclear Fuel Waste Bill, second reading – debate continued.

–. House of Commons. 2002c. *Debates*, 13 June.

Canadian Coalition for Ecology, Ethics, and Religion (CCEER). 1996. *A Report to the FEARO Panel*. Toronto: CCEER.

Canadian Electrical Association (CEA). 2006. *Power Generation in Canada: A Guide*. Ottawa: CEA. http://www.canelect.ca.

Canadian Environmental Assessment Agency (CEAA). 1995a. *Compendium of Public Comments on the Adequacy of the Environmental Impact Statement on the Nuclear Fuel Waste Management and Disposal Concept*. Report of the Nuclear Fuel Waste Disposal Concept Environmental Assessment Panel. 2 vols. Ottawa: Minister of Supply and Services.

–. 1995b. *An Evaluation of the Environmental Impact Statement on Atomic Energy of Canada Limited's Concept for Disposal of Canada's Nuclear Fuel Waste*. Report of the Scientific Review Group, Advisory to the Nuclear Fuel Waste Management and Disposal Concept Environmental Assessment Panel. Ottawa: Minister of Supply and Services.

–. 1996. *Compendium of Written Submissions for Phase I of Public Hearings*. Nuclear Fuel Waste Disposal Concept Environmental Assessment Panel. Vols. 1-6. Hull, QC: Ministry of Supply and Services.

–. 1997. *Nuclear Fuel Waste Management and Disposal Concept Environmental Assessment Panel Public Hearings Transcripts, 1996-1997*. 52 vols. Toronto: Farr and Associates.

–. 1998. *Report of the Nuclear Fuel Waste Management and Disposal Concept Environmental Assessment Panel*. No. EN-106 30/1-1998E. Ottawa: Public Works and Government Services Canada.

Canadian Environmental Assessment Panel (CEAP). 1992. *Final Guidelines for the Preparation of an Environmental Impact Statement on the Nuclear Fuel Waste Management and Disposal Concept.* Ottawa: Public Works and Government Services Canada.

Canadian Nuclear Association (CNA). 1975. *Nuclear Power in Canada: Questions and Answers.* Toronto: CNA.

–. 2007. *Canada's Nuclear Energy: Reliable, Clean, and Affordable Electricity.* Toronto: CNA. http://www.cna.ca.

Canadian Policy Research Networks (CPRN). 2004. *Responsible Action: Citizens' Dialogue on the Long Term Management of Used Nuclear Fuel.* Research Report p/04. Ottawa: CPRN. http://www.nwmo.ca.

Chambers, S. 1996. *Reasonable Democracy: Jürgen Habermas and the Politics of Discourse.* Ithaca: Cornell University Press.

Citizens for Renewable Energy. 2005. *Submission to NWMO National Consultation Process.* Toronto: NWMO.

Coal Industry Advisory Board. 1995. *Regional Trends in Energy-Efficient, Coal-Fired, Power Generation Technologies.* Report produced by the Global Climate Committee of the International Atomic Energy Agency's Coal Industry Advisory Board. http://www.iaea.org.

Cohen, J. 1997a. "Deliberation and Democratic Legitimacy." In *Deliberative Democracy: Essays on Reason and Politics,* ed. J. Bohman and W. Rehg, 67-91. Cambridge, MA: MIT Press.

–. 1997b. "Procedure and Substance in Deliberative Democracy." In *Deliberative Democracy: Essays on Reason and Politics,* ed. J. Bohman and W. Rehg, 407-37. Cambridge, MA: MIT Press.

Coleman, Bright Associates, and Patterson Consulting. 2003. *Development of the Environmental Component of the NWMO Analytical Framework: NWMO Background Papers and Workshop Reports.* Toronto: NWMO.

Community of Inuvik. 2005. *Unikkaaqatigiit: Putting the Human Face on Climate Change – Perspectives from Inuvik, Inuvialuit Settlement Region.* Ottawa: Joint Publication of Inuit Tapiriit Kanatami, Nasivvik Centre for Inuit Health and Changing Environments at Université Laval, and the Ajunnginiq Centre at the National Aboriginal Health Organization.

Congress of Aboriginal Peoples (CAP). 2005. *Summary of Key Observations Regarding NWMO: Discussion Document 2 – Understanding the Choices.* Toronto: NWMO.

COWAM. 2002-3. "Case Study: Tierp and Oskarshamn." http://www.cowam.com/spip.php?article30.

Dales, John. H. 1953. "Fuel, Power, and Industrial Development in Central Canada." *American Economic Review* 43, 2: 181-98.

Daniels, R.J., ed. 1996. *Ontario Hydro at the Millennium: Has Monopoly's Moment Passed?* Montreal: McGill-Queen's University Press.

Daniels, R.J., and M.J. Trebilcock. 1996. "The Future of Ontario Hydro: A Review of Structural and Regulatory Options." In *Ontario Hydro at the Millennium: Has Monopoly's Moment Passed?* ed. R.J. Daniels, 1-52. Montreal: McGill-Queen's University Press.

Denison, M. 1960. *The People's Power: The History of Ontario Hydro.* Toronto: McClelland and Stewart.

Desveaux, J.A., E.A. Lindquist, and G. Toner. 1994. "Organizing for Policy Innovation in Public Bureaucracy: AIDS, Energy, and Environmental Policy in Canada." *Canadian Journal of Political Science* 27, 3: 493-528.

Dewees, Donald N. 2001. "The Future of Nuclear Power in a Restructured Electricity Market." In *Canadian Nuclear Energy Policy: Changing Ideas, Institutions, and Interests,* ed. G. Bruce Doern, Arslan Dorman, and Robert W. Morrison, 147-73. Toronto: University of Toronto Press.

Doern, G.B. 1977. *The Atomic Energy Control Board.* Ottawa: Supply and Services Canada.

Doern, G.B., A. Dorman, and R.W. Morrison. 2001a. "Precarious Opportunity: Canada's Changing Nuclear Energy Policies and Institutional Choices." In *Canadian Nuclear Energy Policy: Changing Ideas, Institutions, and Interests,* ed. G.B. Doern, A. Dorman, and R.W. Morrison, 1-33. Toronto: University of Toronto Press.

–. 2001b. "Transforming AECL into an Export Company: Institutional Challenges and Change." In *Canadian Nuclear Energy Policy: Changing Ideas, Institutions, and Interests,* ed. G.B. Doern, A. Dorman, and R.W. Morrison, 74-95. Toronto: University of Toronto Press.

Doern, G.B., and M. Gattinger. 2003. *Power Switch: Energy Regulatory Governance in the Twenty-First Century.* Toronto: University of Toronto Press.

Doern, G.B., and R.W. Morrison, eds. 1980. "The Politics of Canadian Nuclear Energy." In *Canadian Nuclear Policies,* ed. G.B. Doern and R.W. Morrison, 45-58. Montreal: Institute for Research on Public Policy.

–. 1981. *Government Intervention in Canadian Nuclear Policies.* Toronto: Institute for Public Research on Policy.

Doern, G.B., and V.S. Wilson. 1974. "Conclusions and Observations." In *Issues in Canadian Public Policy,* ed. G.B. Doern and V.S. Wilson, 337-45. Toronto: Macmillan.

Douglas, M. 1991. "Risk Acceptability According to the Social Sciences." In *Communities of Fate: Readings in the Social Organization of Risk,* ed. C.E. Marske, 169-86. New York: University Press of America.

Dowdeswell, E., president of the NWMO. 2005. Interview with Johnson, Toronto, 6 June.

–. 2004b. *Durham Nuclear Health Committee NWMO Dialogue on the Future of Canada's Used Nuclear Fuel.* Toronto: NWMO.

DPRA Canada. 2004a. *Durham Nuclear Health Committee NWMO Dialogue on the Future of Canada's Used Nuclear Fuel.* Toronto: NWMO.

–. 2004b. *Final Report: National Stakeholders and Regional Dialogue Sessions.* Toronto: NWMO.

–. 2004c. *The Future of Canada's Used Nuclear Fuel: International Youth Nuclear Congress Round Table.* Toronto: NWMO.

Duquette, M. 1995. "From Nationalism to Continentalism: Twenty Years of Energy Policy in Canada." *Journal of Socio-Economics* 24, 1: 229-51.

Durant, D. 2006. "Managing Expertise: Performers, Principals, and Problems in Canadian Nuclear Waste Management." *Science and Public Policy* 33, 3: 191-204.

–. 2007a. "Burying Globally, Acting Locally: Control and Co-Option in Nuclear Waste Management." *Science and Public Policy* 34, 7: 515-28.

–. 2007b. "Resistance to Nuclear Waste Disposal: Credentialed Experts, Public Opposition, and Their Shared Lines of Critique." *Scientia Canadensis* 30, 1: 1-30.

–. 2009a. "Radwaste in Canada: A Political-Economy of Uncertainty." *Journal of Risk Research* 31, 4.

–. 2009b. "Responsible Action and Nuclear Waste Disposal." *Technology in Society* 31, 2: forthcoming.

Edwards, G. 2005. *Following the Path Backwards: A Critique of the Draft Study Report of the NWMO Entitled "Choosing a Way Forward" (May 2005).* Report for the Canadian Coalition for Nuclear Responsibility. http://www.ccnr.org.

Edwards, G., and R. Del Tredici. 1999. *The Atomic Atlas of Canada: Overview and Map.* Ottawa: Radioactive Inventory Project. Available from Campaign for Nuclear Phaseout.

Eggleston, W. 1965. *Canada's Nuclear Story.* Toronto: Clark-Irwin.

Elam, M., and G. Sundqvist. 2007. *Swedish Involvement in Swedish Nuclear Waste Management.* SKI Report 2007: 02. http://www.ski.se/dynamaster/file_archive/.

Energy, Mines, and Resources (EMR) Canada. 1982. *Nuclear Policy Review Background Papers.* Report ER 81-2E. Ottawa: EMR.

–. 1988a. *AECL and the Future of the Canadian Nuclear Industry: A Proposed Discussion Paper.* Ottawa: EMR.

–. 1988b. *Response of the Government of Canada to the Report of the Standing Committee on Environment and Forestry, "High-Level Radioactive Waste in Canada: The Eleventh Hour."* Ottawa: EMR.

Energy, Mines, and Resources (EMR) Canada and Canadian Electricity Association (CEA). 1992. *Electrical Power in Canada.* Ottawa: Ministry of Supply and Services.

Energy, Mines, and Resources (EMR) Minister and Ontario Energy Minister (OEM). 1978. *Canada/Ontario Joint Statement on the Radioactive Waste Management Programme.* Ottawa: Ministry of Supply and Services.

–. 1981. *Canada/Ontario Joint Statement on the Radioactive Waste Management Programme.* Ottawa: Ministry of Supply and Services.

Ernst and Young. 1993. *The Economic Effects of the Canadian Nuclear Industry.* Report commissioned by AECL. Chalk River: AECL.

Federal Environmental Assessment Review Office (FEARO). 1993. Federal Environmental Assessment Review Panel on the Decommissioning of the Uranium Mine Tailings Management Areas in Elliot Lake, Ontario. Transcript of scoping sessions held in the Serpent River First Nation, Ontario, 15 December. Toronto: International Rose Reporting.

–. 1996. Federal Environmental Assessment Review Panel on the Decommissioning of the Uranium Mine Tailings Management Areas in Elliot Lake, Ontario. Transcript of public hearing sessions held in the Serpent River First Nation, Ontario 23 January. Toronto: International Rose Reporting.

Finch, R. 1986. *Exporting Danger: A History of the Canadian Nuclear Energy Export Programme.* Montreal: Black Rose Books.

Fiorino, D.J. 1990. "Citizen Participation and Environmental Risk: A Survey of Institutional Mechanisms." *Science, Technology, and Human Values* 15, 2: 226-43.

Flynn, J., J. Chalmers, D. Easterling, R. Kasperson, H. Kunreuther, C.K. Mertz, A. Mushkatel, K.D. Pijawaka, and P. Slovic. 1995. *One Hundred Centuries of Solitude: Redirecting America's High-Level Nuclear Waste Policy.* Boulder: Westview Press.

Foucault, M. 1975. *Surveiller et Punir.* Paris: Gallimard.

Freeman, N.B. 1996. *The Politics of Power: Ontario Hydro and Its Government, 1906-1995.* Toronto: University of Toronto Press.

Freeman, S. 2000. "Deliberative Democracy: A Sympathetic Comment." *Philosophy and Public Affairs* 29: 371-418.

Freudenburg, W.R. 2004. "Can We Learn from Failure? Examining U.S. Experiences with Nuclear Repository Siting." *Journal of Risk Research* 7, 2: 153-69.

Gerrard, M.B. 1994. *Whose Backyard, Whose Risk: Fear and Fairness in Toxic and Nuclear Waste Siting.* London: MIT Press.

Giddens, A. 1984. *The Constitution of Society.* Berkeley: University of California Press.

–. 1990. *The Consequences of Modernity.* Cambridge, UK: Polity.

–. 1994. "Living in a Post-Traditional Society." In *Reflexive Modernization: Politics, Tradition, and Aesthetics in the Modern Social Order,* ed. U. Beck, A. Giddens, and S. Lash, 56-109. Stanford: Stanford University Press.

Global Business Network (GBN). 2003. *Looking Forward to Learn: Future Scenarios for Testing Different Approaches to Managing Used Nuclear Fuel in Canada.* Toronto: NWMO.

Government of Canada. 1998. *Response to Recommendations of the Nuclear Fuel Waste Management and Disposal Concept Environmental Assessment Panel.* Ottawa: Queen's Printer. http://rncan.gc.ca/eneene/sources/uranuc/wasdec/nfwdcn/govgou-eng.php.

–. 2002. Nuclear Fuel Waste Management Act (Bill C-27). *Statutes of Canada 49-50-51,* c. 23.

Gray, J.L. 1987. "Early Decisions in the Development of the CANDU Program." *Nuclear Journal of Canada* 1, 2: http://www.cns-snc.ca.

Greber, M.A., E.R. French, and J.A.R. Hillier. 1994. *The Disposal of Canada's Nuclear Fuel Waste: Public Involvement and Social Aspects (R-Public).* AECL Report 10712, COG-93-2. Pinawa: AECL.

Gutmann, A., and D. Thompson. 1999. "Reply to Critics: Democratic Disagreement." In *Deliberative Politics,* ed. S. Macedo, 243-79. New York: Oxford University Press.

–. 2000. *Democracy and Disagreement.* Cambridge, MA: Belknap Press.

Habermas, J. 1973. *Legitimation crisis.* Trans. T. McCarthy. Boston: Beacon Press.

–. 1990. *Moral consciousness and communicative action.* Trans. C. Lenhardt and S. Weber Nicholson. Cambridge: MIT Press.

Halstead, R., F. Dilger, and J.D. Ballard. 2004. "Beyond the Mountains: Nuclear Waste Transportation and the Rediscovery of Nevada." Paper presented at the Waste Management Conference, Tucson, 29 February-4 March. http://www.state.nv.us/nucwaste/news2004/pdf/wm0304.pdf.

Hampton, H. 2003. *Public Power: Energy Production in the 21st Century.* Toronto: Insomniac Press.

Hardy, D., principal of Hardy, Stevenson, and Associates Limited. 2001. Interview with Johnson, Toronto, 26 October.

–. 2005. Interview with Johnson, Toronto, 30 September.

Hardy, Stevenson, and Associates. 1991. *Moral and Ethical Issues Related to the Nuclear Fuel Waste Concept.* AECL Technical Report TR-549, COG-91-140. Toronto: Hardy, Stevenson, and Associates.

–. 1993. *Nuclear Fuel Waste Management Concept: Literature Review and Analysis – Moral and Ethical Issues.* Toronto: Hardy, Stevenson, and Associates.

–. 2005a. *Final Report: National Stakeholders and Regional Dialogue Sessions.* Toronto: NWMO.

–. 2005b. *NWMO Community Dialogue Workshop on Discussion Document Two: Final Report.* Toronto: NWMO.

Hare, F.K. 1997. Testimony of 20 June 1996, in *Nuclear Fuel Waste Management and Disposal Concept Environmental Assessment Panel Public Hearings Transcripts, 1996-1997,* vol. 23, CEAA, 54-87. Toronto: Farr and Associates.

–. 1998. *Independent Review Panel on Nuclear Waste Strategies and Management Processes.* Ontario Hydro report, 30 May.

Harley, M.L. 2004. *What "Newness" Can the Present Process Bring to the Nuclear Fuel Waste Issues?* Toronto: NWMO. http://www.nwmo.ca.

–, member of the Nuclear Wastes Writing Team, United Church of Canada. 2005. Response to an e-mail questionnaire from Johnson, 4 August.

HBT AGRA Limited. 1993. *Aboriginal Involvement in the Nuclear Industry: Case Studies.* AECL research, Pinawa, Whiteshell Laboratories, TR-M 38, April.

Holmstrand, O. 1999. Participation of Local Citizens' Groups in the Swedish Nuclear Waste Process. Paper presented at the Values in Decisions on Risk (VALDOR) Conference, Stockholm, May.

–. 2003. "Nuclear Waste Management in Sweden in Comparison with Other European Countries: NGO Experiences of the COWAM Process." Paper presented at the Values in Decisions on Risk (VALDOR) Conference, Stockholm, May.

Hunold, C. 2002. "Canada's Low-Level Radioactive Waste Disposal Problem: Voluntarism Revisited." *Environmental Politics* 11, 2: 49-72.

Iacobucci, E., M. Trebilcock, and R.A. Winter. 2006. "The Canadian Experience with Deregulation." *University of Toronto Law Journal* 56, 1: 1-74.

Inglehart, R. 1999. "Postmodernism Erodes Respect for Authority, but Increases Support for Democracy." In *Critical Citizens: Global Support for Democratic Governance,* ed. P. Norris, 236-56. Oxford: Oxford University Press.

Institute for Research on Environment and Economy (IREE). 1995. *An Environmental Perspective on Canadian Policy for Sustainable Development.* Ottawa: IREE, University of Ottawa.

International Institute of Concern for Public Health (IICPH). 2007. "Burning Radioactive Waste." Newsletter, spring. http://www.iicph.org/docs/.

Inuit Tapiriit Kanatami (ITK). 2005. *Final Report on the National Inuit Specific Dialogues on the Long-Term Management of Nuclear Fuel Waste in Canada.* Ottawa. http://www.nwmo.ca.

Jaccard, M. 1995. "Oscillating Currents: The Changing Rationale for Government Intervention in the Electricity Industry." *Energy Policy* 23, 7: 579-92.

Jackson, David, and John de la Mothe. 2001. "Nuclear Regulation in Transition: The Atomic Energy Control Board." In *Canadian Nuclear Energy Policy,* eds. G. B. Doern, A. Dorman, and R.W. Morrison, 96-112. Toronto: University of Toronto Press.

Jackson, J. 2005. Letter from Citizens' Network on Waste Management to NWMO, 31 August. http://www.nwmo.ca.

Jacob, G. 1990. *Site Unseen: The Politics of Siting a Nuclear Waste Repository.* Pittsburgh: University of Pittsburgh Press.

Janes, G. (National Council of Women of Canada). 2005. Comments on the NWMO Draft Management Plan for Nuclear Waste. http://www.nwmo.ca.

Jenkins, G.P. 1985. "Public Utility Finance and Economic Waste." *Canadian Journal of Economics* 18, 3: 484-98.

Johannson, P.R., and J.C. Thomas. 1981. "A Dilemma of Nuclear Regulation in Canada: Public Control and Public Confidence." *Canadian Public Policy* 7, 3: 433-43.

Johnson, G.F. 2007. "The Discourse of Democracy in Canadian Nuclear Waste Management Policy." *Policy Sciences* 40: 79-99.

–. 2008. *Deliberative Democracy for the Future: The Case of Nuclear Waste Management in Canada.* Toronto: University of Toronto Press.

Jonas, H. 1984. *The Imperative of Responsibility: In Search of an Ethics for the Technological Age.* Chicago: University of Chicago Press.

Joppke, C. 1992-93. "Decentralization of Control in U.S. Nuclear Energy Policy." *Political Science Quarterly* 107, 4: 709-25.

Kamps, K. 2005. *Submission of Comments to Canadian Nuclear Waste Management Organization Regarding Its "Choosing a Way Forward: Draft Study Report."* Washington, DC: Nuclear Information and Resource Service.

Kneen, J. 2006. "Uranium Mining in Canada: Past and Present." Notes for a presentation to the Indigenous World Uranium Summit, 30 November-1 December. http://www.miningwatch.ca/updir/Uranium_Canada_web.pdf.

Kneen, S., co-ordinator of the National Inuit Specific Dialogue, Environment Department, Inuit Tapiriit Kanatami. 2005. Interview with Johnson, Ottawa, 9 May.

Knelman, F. 1976. *Nuclear Energy: The Unforgiving Technology.* Edmonton: Hurtig Publishers.

Kraft, M.E., and B.B. Clary. 1991. "Citizen Participation and the NIMBY Syndrome: Public Response to Radioactive Waste Disposal." *Western Political Quarterly* 44, 2: 299-328.

Kuhn, R.G. 1997. "Public Participation in the Hearings on the Canadian Nuclear Fuel Waste Disposal Concept." In *Canadian Assessment in Transition,* ed. A.J. Sinclair, 19-49. Waterloo: Department of Geography, University of Waterloo.

Kuhn, R.G., and B.L. Murphy. 2003. "Economic Regions." NWMO background document 5-1. http://www.nwmo.ca/.

–. 2006. "Environmental Justice, Place, and Nuclear Fuel Waste Management in Canada." Paper presented at the Values in Decisions on Risk (VALDOR) Conference, Stockholm, May.

Kunreuther, H., K. Fitzgerald, and T.D. Aarts. 1993. "Siting Noxious Facilities: A Test of the Facility Siting Credo." *Risk Analysis* 13, 3: 301-18.

Lake, R.W., and L. Disch. 1992. "Structural Constraints and Pluralist Contradictions in Hazardous Waste Regulation." *Environment and Planning A* 24, 3: 663-81.

Lifton, R.A. 1982. *Indefensible Weapons: The Political and Psychological Case against Nuclearism.* Oxford: Oxford University Press.

Lovins, A. 1976. "Energy Strategy: The Road Not Taken?" *Foreign Affairs* 55, 1: 65-96.

Luhmann, N. 1979. *Trust and Power.* Chichester: John Wiley and Sons.

Lyon, V. 1984. "Minority Government in Ontario, 1975-81: An Assessment." *Canadian Journal of Political Science* 17, 4: 685-705.

Macey, D. 2000. "Rhetoric." In *Dictionary of Critical Theory,* 329-30. London: Penguin Books.

MacIntyre, A. 1984. *After Virtue: A Study in Moral Theory.* 2nd ed. Notre Dame, IN: University of Notre Dame Press.

Manson, N. 2002. "Formulating the Precautionary Principle." *Environmental Ethics* 24: 263-74.

Martin, D.H. 2000. *Financial Meltdown: Federal Nuclear Subsidies to AECL.* Ottawa: Campaign for Nuclear Phaseout. http://www.cnp.ca.

–. 2003. *Canadian Nuclear Subsidies: Fifty Years of Futile Funding, 1952-2002.* Ottawa: Campaign for Nuclear Phaseout. http://www.cnp.ca.

Martin, D.H., and D. Argue. 1996. *Nuclear Sunset: The Economic Costs of the Canadian Nuclear Industry.* Report produced for the Campaign for Nuclear Phaseout. http://www.ccnr.org.

Mawhinney, H.B. 2001. "Theoretical Approaches to Understanding Interest Groups." *Educational Policy* 15, 1: 187-214.

McCloskey, J. 1995. "Briefing (June 5, 1995) to the Minister of Natural Resources Canada Regarding the Natural Resources Canada Stakeholder Consultations for the Development of a Federal Policy Framework on Radioactive Waste." Document obtained under federal access to information legislation.

McCombie, C., and B. Tveiten. 2004. "A Comparative Overview of Approaches to Management of Spent Nuclear Fuel and High Level Wastes in Different Countries." NWMO background document 7-6. http://www.nwmo.ca/.

McCool, D. 1998. "The Subsystem Family of Concepts: A Critique and a Proposal." *Political Research Quarterly* 51, 2: 551-70.

McDougall, J.N. 1982. *Fuels and the National Policy.* Toronto: Butterworths.

McKay, P. 1983. *Electric Empire: The Inside Story of Ontario Hydro.* Toronto: Between the Lines.

McMurdy, D. 1993. "Maurice Strong Restructures Ontario Hydro." *Maclean's,* April 26, 18-19.

Means, E. 2007. "Trust in the Case of Port Credit." PhD diss., York University.

Mehta, M.D. 2003. "Energy Mixes and Future Scenarios: The Nuclear Option Deconstructed." In *The Integrity Gap: Canada's Environmental Policy and Institutions,* ed. E. Lee and A. Perl, 105-32. Vancouver: UBC Press.

–. 2005. *Risky Business: Nuclear Power and Public Protest in Canada.* Lanham, MD: Lexington Books.

Member, NWMO Advisory Council. 2005. Telephone interview with Johnson, 5 October.

Métis Nation of Ontario. 2008. "Towards a Consultation Framework for Ontario Métis." Ottawa: Métis Nation of Ontario.

Miller, B. 1992. "Collective Action and Rational Choice: Place, Community, and the Limits to Individual Self Interest." *Economic Geography* 68: 22-42.

Morgan, W., ed. 1977. *Report by the Committee Assessing Fuel Storage.* AECL Report 5959/1. Ottawa: AECL.

Morrison, R.W. 1998. *Nuclear Energy Policy in Canada 1942-97.* Report commissioned by NRCan and written for the Carleton Research Unit on Innovation Science and Environment (CRUISE).

Murphy, B.L. 2001. "Canadian Nuclear Fuel Waste: Current Contexts and Future Management Prospects." PhD diss., University of Guelph.

–. 2003. "An Evaluation of the Canadian Nuclear Fuel Waste Management Act." In *The 10th International High-Level Radioactive Waste Management Conference,* ed. American Nuclear Society, 1193-99. Las Vegas, NV: American Nuclear Society.

Murphy, B.L., and R.G. Kuhn. 2001. "Setting the Terms of Reference in Environmental Assessments: Canadian Nuclear Fuel Waste Management." *Canadian Public Policy* 27, 3: 249-66.

–. 2006. "Scaling Environment Justice: The Case of the Waste Isolation Pilot Plant." Paper presented at the Values in Decisions on Risk (VALDOR) Conference, Stockholm, May. http://www.congrex.com/valdor2006/papers/26_Murphy.pdf.

Murphy, B.L., and A. Stanley. 2006. "The Voluntary Siting Model: Implications for Environmental Justice." In *The 11th International High-Level Radioactive Waste Management Conference,* ed. American Nuclear Society, 1180-86. Las Vegas: American Nuclear Society.

Musolf, L.D. 1956. "Canadian Public Enterprise: A Character Study." *American Political Science Review* 50, 2: 405-21.

Mutton, J., mayor of Clarington and chair of Canadian Association of Nuclear Host Communities. 2005. Interview with Johnson, Clarington, 7 June.

Nagra. 2006. "Feasibility of Safe Deep Disposal of High-Level Waste in Switzerland Confirmed." http://www.nagra.ch/.

Native Women's Association of Canada (NWAC). 2005. "National Consultation on Nuclear Fuel Waste Management." Ottawa, 14 June. http://www.nwmo.ca.

Natural Resources Canada (NRCan). 1995. "The Development of a Federal Policy Framework for the Disposal of Radioactive Wastes in Canada." Document 000795. Obtained under federal access to information legislation.

–. 1996. *Policy Framework for Radioactive Waste.* Ottawa: Government of Canada.

–. 1998. *Government of Canada Response to Recommendations of the Nuclear Fuel Waste Management and Disposal Concept Environmental Assessment Panel.* Ottawa: Ministry of Supply and Services.

–. 2005. "Notes for a Speech by John Efford, Minister of NRCan, to the CNA Annual Seminar, 9 March 2005." http://www.nrcan.gc.ca/media/speeches/2005/200519_e.htm.

–. 2007. "Canada's Nuclear Future: Clean, Safe, Responsible." Press release, 14 June. http://www.nrcan.gc.ca/media.

Norris, P. 1999. "Introduction: The Growth of Critical Citizens." In *Critical Citizens: Global Support for Democratic Governance,* ed. P. Norris, 1-24. Oxford: Oxford University Press.

Northwatch. 2004. "Nuclear Waste Debate Heats Up across Northern Ontario." *Northwatch News,* summer, 1.

–. 2006. "North Shore Threatened with Nuclear Expansion." *Northwatch News,* spring, 1.

Nuclear Energy Agency (NEA). 2001. *Annual Report.* Organization for Economic Co-Operation and Development (OECD) Publications 81767 2002. Paris: OECD.

Nuclear industry spokesperson. 2005. Telephone interview with Johnson, 20 October.

Nuclear Waste Management Organization (NWMO). 2003. *Asking the Right Questions? The Future Management of Canada's Used Nuclear Fuel.* Discussion Document 1. Toronto: NWMO.

–. 2004a. *Assessing the Options: Future Management of Used Nuclear Fuel in Canada.* NWMO Assessment Team report. Toronto: NWMO.

–. 2004b. *From Dialogue to Decision: Managing Canada's Nuclear Fuel Waste.* Annual report. Toronto: NWMO.

–. 2004c. Information meeting notes. Toronto: NWMO.

–. 2004d. *Roundtable on Ethics: Ethical and Social Framework.* NWMO Background Papers and Workshop Reports. Toronto: NWMO.

–. 2004e. *Understanding the Choices: The Future Management of Canada's Used Nuclear Fuel.* Discussion Document 2. Toronto: NWMO.

–. 2005a. *Adaptive Phased Management: Technical Description.* NWMO Background Paper 6-18. Toronto: NWMO.

–. 2005b. *Choosing a Way Forward: The Future Management of Canada's Used Nuclear Fuel –Draft Study Report.* Toronto: NWMO.

–. 2005c. *Choosing a Way Forward: The Future Management of Canada's Used Nuclear Fuel –Final Study.* Toronto: NWMO.

–. 2005d. *Incorporation of Seaborn Panel Recommendations and Insights in the Work of the NWMO.* NWMO Background Papers: Social and Ethical Dimensions, 2-8. Toronto: NWMO.

–. 2006. *Looking Ahead, Planning for the Future.* Annual report. http://www.nwmo.ca/.

Nuclear Waste Watch. 2004. *Position Statement.* Toronto: Nuclear Waste Watch.

Office of the Auditor General of Canada (OAG). 2000. *Canadian Nuclear Safety Commission: Power Reactor Regulation.* Chapter 27 of the 2000 reports of the OAG. http://www.oag-bvg.gc.ca.

–. 2005. *Canadian Nuclear Safety Commission: Power Reactor Regulation.* Chapter 6 of the 2005 reports of the OAG. http://www.oag bvg.gc.ca.

Oliga, J.C. 1996. *Power, Ideology, and Control.* New York: Plenum Press.

Ontario Power Authority (OPA). 2005. *Supply Mix Advice Report.* Vol. 1. http://www.power-authority.on.ca.

–. 2007. *Ontario's Integrated Power System Plan: The Road Map for Ontario's Electricity Future.* http://www.powerauthority.on.

Ontario Power Generation (OPG). 2003. *Response to Technical Comments Raised during Environmental Assessment of AECL Disposal Concept.* OPG document in support of *Background Paper 6-9: Conceptual Designs for Used Nuclear Fuel Management.* Toronto: NWMO.

–. 2006. "Ontario Power Generation Begins Federal Approvals Process for Potential New Nuclear Units." Press release, 22 September. http://www.opg.com.

Openshaw, S., S. Carver, and J. Fernie. 1989. *Britain's Nuclear Waste: Safety and Siting.* London: Belhaven Press.

Oskarshamn. 2003. "Nuclear Waste: Local Competence Building in Oskarshamn." http://web.wpab.se/lko/.

Otway, H. 1992. "Expert Fallibility, Public Wisdom: Toward a Contextual Theory of Risk." In *Social Theories of Risk,* ed. S. Krimsky and D. Golding, 215-28. New York: Praeger.

Oxman, A.D., H.S. Shannon, W.J. Garland, and G.W. Torrance. 1989. "Nuclear Safety in Ontario: A Critical Review of Quantitative Analysis." *Risk Analysis* 9, 1: 43-54.

Parfit, D. 1984. *Reasons and Persons.* Oxford: Oxford University Press.

Parliament of Canada (Thirty-Seventh). 2002. *An Act Respecting the Long-Term Management of Nuclear Fuel Waste.* Ottawa: Queen's Printer.

Parsons, G.F., and R. Barsi. 2001. "Uranium Mining in Northern Saskatchewan: A Public-Private Transition (Part 1)." In *Large Mines and the Community: Socioeconomic and Environmental Effects in Latin America, Canada, and Spain,* ed. G. McMahon and F. Remy. International Development Research Centre Books. Also available at http://www.idrc.ca/fr/ev-28034-201-1-DO_TOPIC.html.

Pembina Institute. 2004. *Power for the Future: Towards a Sustainable Electricity System.* http://www.pembina.org.

–. 2006. *Nuclear Power in Canada: An Examination of Risks, Impacts, and Sustainability.* http://www.pembina.org.

Pijawka, K.D., and A.H. Mushkatel. 1992. "Public Opposition to the Siting of the High-Level Nuclear Waste Repository: The Importance of Trust." *Policy Studies Review* 10, 4: 180-94.

Poch, D. 1986. "Acting Locally." In *Challenges to Nuclear Waste: Proceedings of the Nuclear Waste Issues Conference,* ed. A. Wieser, 206-11. Winnipeg: Lac Du Bonnet.

Porter, A., chairman, Royal Commission on Electric Power Planning. 1978. *A Race against Time: Interim Report on Nuclear Power in Ontario.* Toronto: Queen's Printer for Ontario.

Prichard, J.R.S, and M.J. Trebilcock. 1983. "Crown Corporations in Canada: The Choice of Instrument." In *The Politics of Canadian Public Policy,* ed. M.M. Atkinson and M.A. Chandler, 199-222. Toronto: University of Toronto Press.

Pross, A.P. 1985. "Parliamentary Influence and the Diffusion of Power." *Canadian Journal of Political Science* 18, 2: 235-66.

Rayner, S., and R. Cantor. 1987. "How Fair Is Safe Enough: The Cultural Approach to Technology Choice." *Risk Analysis: An International Journal* 7, 1: 3-9.

Reckmans, L., K. Lewis, and A. Dwyer, eds. 2003. *This Is My Homeland: Stories of the Effects of Nuclear Industries by People of the Serpent River First Nation and the North Shore of Lake Huron.* Serpent River First Nation: Anishnabe Printing.

Reddy, S.G. 1996. "Claims to Expert Knowledge and the Subversion of Democracy: The Triumph of Risk over Uncertainty." *Economy and Society* 25: 222-54.

Rees, J. 1985. *Natural Resources: Allocation, Economics, and Policy.* London: Methuen.

Robbins, W. 1984. *Getting the Shaft: The Radioactive Waste Controversy in Manitoba.* Winnipeg: Lac Du Bonnet.

–. 1986. "Citizen Advocacy and Action." In *Challenges to Nuclear Waste: Proceedings of the Nuclear Waste Issues Conference,* ed. A. Wieser, 170-76. Winnipeg: Lac Du Bonnet.

Robertson, J.A.L. 1998. "Malice in Blunderland?" *Canadian Nuclear Society Bulletin* 19, 2-3. http://www.cns-snc.ca.

Romerio, F. 1998. "The Risks of Nuclear Policies." *Energy Policy* 26, 3: 239-46.

Rubin, N., director of nuclear research and senior policy analyst, Energy Probe. 2001. Interview with Johnson, Toronto, 1 August.

Salaff, S. 2005. "Native Communities Refuse Nuclear Waste." *Seven Oaks Magazine,* 27 October. http://www.sevenoaks.com.

Salter, L., and D. Slaco. 1981. "Environmental Assessment at Point Lepreau." In *Public Inquiries in Canada,* ed. L. Salter and D. Slaco, 47-62. Ottawa: Canadian Government Publishing Centre.

Saskatchewan Mining Association. "Commodities Mined in Saskatchewan." http://www.saskmining.ca/.

Schively, C. 2007. "Understanding the NIMBY and LULU Phenomena: Reassessing Our Knowledge Base and Informing Future Research." *Journal of Planning Literature* 21, 3: 255-66.

Schrecker, T. 1987. "The Atomic Energy Control Board: Assessing Its Role in Reactor Safety Regulation." Submission to the Ontario Nuclear Safety Review on behalf of Energy Probe, 1 September. Document EP-307. http://www.energyprobe.org.

Select Committee on Ontario Hydro Affairs. 1980. *The Management of Nuclear Fuel Waste: Final Report.* Submitted to the Legislative Assembly of Ontario, Third and Fourth Sessions, Thirty-First Parliament. Toronto: Select Committee.

Sharma, N. 2006. *Home Economics: Nationalism and the Making of Migrant Workers in Canada.* Toronto: University of Toronto Press.

Shoesmith, D. 2005. Telephone interview with Johnson, 15 September.

Shoesmith, D., and L. Shemilt. 2003. *Workshop on the Technical Aspects of Nuclear Fuel Waste Management: Executive Summary.* NWMO Background Papers and Workshop Reports. Toronto: NWMO.

Shrader-Frechette, K. 2003. *Risk and Uncertainty.* Toronto: NWMO. http://www.nwmo.ca.

Sierra Club of Canada. 2001. *Financing Disaster: How the G8 Fund the Global Proliferation of Nuclear Technology.* Uxbridge, ON: Sierra Club of Canada Nuclear Campaign. http://www.sierraclub.ca.

Sierra Legal Defense. 2006. *Application for Inquiry: Filed Pursuant to Section 9 of the Competition Act against the Canadian Nuclear Society.* http://www.sierralegal.org.

Sigurdson, G., and B. Stuart. 2003. *A Planning Workshop.* NWMO Background Papers and Workshop Reports. Toronto: NWMO.

Silk, J. 1999. "Guest Editorial: The Dynamics of Community, Place, and Identity." *Environment and Planning A* 31: 5-17.

Simpson, L., mayor of Pinawa. 2005. Telephone interview with Johnson, Montreal, 23 September.

Sims, G.H.E. 1981. *A History of the Atomic Energy Control Board.* Ottawa: Minister of Supply and Services Canada.

Slovic, P. 1987. "Perceptions of Risk." *Science* 236, 4799: 280-85.

–. 1992. "Perception of Risk: Reflections on the Psychometric Paradigm." In *Social Theories of Risk,* ed. S. Krimsky and D. Golding, 117-52. New York: Praeger.

Soderholm, P. 1998. "Fuel Choice in West European Power Generation since the 1960s." *OPEC Review* 22, 3: 201-31.

Solomon, L. 1984. *Power at What Cost? Why Ontario Hydro Is out of Control and What Needs to Be Done about It.* Toronto: Energy Probe Research Foundation.

Staff, NWMO. 2005. Interview with Johnson, Toronto, 6 June and 30 September.

Staff #One, Assembly of First Nations (AFN). 2001. Interview with Johnson, Ottawa, 23 August.

Staff #Two, AFN. 2005. Interview with Johnson, Ottawa, 13 October.

Standing Committee on Energy, Mines, and Resources. 1988. *Nuclear Energy: Unmasking the Mystery.* Second Session of the Thirty-Third Parliament Ottawa: Ministry of Supply and Services.

Standing Committee on Government Agencies. 1992. *Hansard,* Legislative Assembly of Ontario. Testimony of Maurice Strong, 9 December 1992. Available at www.ontla.on.ca.

Stanley, A. 2004. *Summary and Analysis of the Testimonies and Submissions of Aboriginal Peoples and Nations to the Public Environmental Assessment Panel on the Concept of Deep Geologic Disposal in the Canadian Shield.* Report prepared for the AFN. Available from the AFN.

–. 2005. Interview with Johnson, Toronto, 29 September.

–. 2006. "Risk, Scale, and Exclusion in Canadian Nuclear Fuel Waste Management." *ACME: An International E Journal for Critical Geographies* 4, 2: 194-227.

–. 2008. "Citizenship and the Production of Landscape and Knowledge in Contemporary Canadian Nuclear Fuel Waste Management." *Canadian Geographer* 52, 1: 64-82.

Stanley, A., R. Kuhn, and B. Murphy. 2004a. *Response to NWMO:* Asking the Right Questions? Toronto: NWMO.

–. 2004b. *Response to NWMO Discussion Document Two:* Understanding the Choices. Toronto: NWMO.

–. 2005. *Response to NWMO's Draft Study Report:* Choosing a Way Forward. Submission to the NWMO's national consultation process. Toronto: NWMO.

Stevenson, Mark. 2003. "Social and Ethical Issues." NWMO Background Report, 2-8. Toronto: NWMO. Available at http://www.nwmo.ca.

Stratos 2005. Dialogue on *Choosing a Way Forward: The NWMO Draft Study Report – Summary Report.* Toronto: NWMO.

Sundqvist, G. 2002. *The Bedrock of Opinion: Science, Technology, and Society in the Siting of High-Level Nuclear Waste.* Dordrecht: Kluwer Academic.

Sunstein, C. 1997. "Deliberation, Democracy, Disagreement." In *Justice and Democracy: Cross-Cultural Perspectives,* ed. R. Bontekoe and M. Stepaniants, 93-117. Honolulu: University of Hawaii Press.

Surrey, J., and C. Huggett. 1976. "Opposition to Nuclear Power: A Review of International Experience." *Energy Policy* 4: 286-307.

Swift, J., and K. Stewart. 2004. *Hydro: The Decline and Fall of Ontario's Electric Empire.* Toronto: Between the Lines.

–. 2005. "Union Power: The Charged Politics of Electricity in Ontario." *Just Labour* 5: 14-22.

Thomas, S. 2004. "The Ontario Government's Proposals on Electricity Restructuring: Comments by Public Service International Research Unity [PSIRU]." London: PSIRU. http://www.psiru.org.

–. 2005. *The Economics of Nuclear Power: Analysis of Recent Studies.* Prepared for PSIRU. London: PSIRU. http://www.psiru.org.

Tornqvist, M. 1999. "The Myth of the 'Safe Swedish Solution' to the Problem of Nuclear Waste." World Information Service on Energy (WISE) News Communique 515. http://www.folkkampanjen.se/doc1/mt9908.html.

Trebilcock, M., and R. Hrab. 2005. "Electricity Restructuring in Ontario." *Energy Journal* 26, 1: 123-46.

Trudeau Foundation and Sierra Club of Canada. 2005. *Roundtable Discussion on Nuclear Waste Management: Proceedings, Response, and Recommendations.* Toronto: NWMO.

United Church of Canada. 1996. *Submission to the Public Hearings of the Canadian Environmental Assessment Panel.* Toronto: United Church, Program Unit on Peace, Environment, and Rural Life.

–. 2004. *Submission One: United Church of Canada General Comments on Nuclear Wastes and the Work of the Nuclear Waste Management Organization.* Toronto: NWMO.

–. 2005a. *The Response of the United Church of Canada to the Nuclear Waste Management Organization Draft Report.* Toronto: NWMO.

–. 2005b. *The Response of the United Church of Canada to the Nuclear Waste Management Organization Final Report.* Toronto: NWMO.

–. 2005c. *Submission Two: Commentary on a United Church of Canada Ethical Lens for Viewing the Problem of Nuclear Wastes.* Toronto: NWMO.

Valadez, J. 2001. *Deliberative Democracy, Political Legitimacy, and Self-Determination in Multicultural Societies.* Boulder: Westview Press.

Watling, J., assistant director, Canadian Policy Research Networks (CPRN). 2005. Interview with Johnson, Ottawa, 9 May.

Watling, J., J. Maxwell, N. Saxena, and S. Taschereau. 2004. *Responsible Action: Citizens' Dialogue on the Long-Term Management of Used Nuclear Fuel.* Public Involvement Network Research Report P/04. Ottawa: CPRN.

Weller, J. 1990. "The Canadian Nuclear Association Turns 30: An Overview of CNA History Spanning Three Decades." *Nuclear Canada: Yearbook 1990.* Toronto: CNA, 9-16.

Welsh, I. 2000. *Mobilizing Modernity: The Nuclear Moment.* London: Routledge.

Wilson, Lois. 2000. *Nuclear Waste: Exploring the Ethical Dilemmas.* Toronto: United Church Publication House.

Winfield, M. 1994. "The Ultimate Horizontal Issue: The Environmental Policy Experiences of Alberta and Ontario, 1971-1993." *Canadian Journal of Political Science* 27, 1: 129-52.

Wolsink, M. 1994. "Entanglement of Interests and Motives: Assumptions behind the NIMBY-Theory on Facility Siting." *Urban Studies* 31, 6: 851-66.

Wynne, B. 1992. "Risk and Social Learning: Reification to Engagement." In *Social Theories of Risk,* ed. S. Krimsky and D. Golding, 275-97. New York: Praeger.

–. 2003. "Seasick on the Third Wave: Subverting the Hegemony of Propositionalism." *Social Studies of Science* 33, 3: 401-17.

–. 2005. "Risk as Globalizing 'Democratic' Discourse? Framing Subjects and Citizens." In *Science and Citizens: Globalization and the Challenges of Engagement,* ed. M. Leach, I. Scoones, and B. Wynne, 66-82. London: Zed Books.

Young, I.M. 1990. *Justice and the Politics of Difference.* Princeton: Princeton University Press.

–. 1999. "Justice, Inclusion, and Deliberative Democracy." In *Deliberative Politics,* ed. S. Macedo, 151-58. New York: Oxford University Press.

Contributors

Darrin Durant is an assistant professor in the Program in Science and Technology Studies at York University, Canada. Dr. Durant is interested in public policy conflicts over issues involving technical claims, especially conflicting expert claims and disputes between experts, governing institutions, and lay public groups. His research focuses on how nuclear waste management, as a technical field of inquiry, is tied up with competing mobilizations of future socio-political and scientific orders. Dr. Durant works closely with scholars from other nations that are also facing the problem of managing nuclear waste and is thus interested in the global dimension of nuclear waste as a part of energy policy on the international stage. He has published numerous scholarly articles and is completing a manuscript on the Canadian debate over nuclear waste management and energy policy.

Genevieve Fuji Johnson is an assistant professor in the Department of Political Science at Simon Fraser University. Dr. Johnson is interested in theories and practices of deliberative democracy, ethics and public policy, and non-positivist policy analysis. She is especially interested in policy areas associated with risk and uncertainty. She is the author of *Deliberative Democracy for the Future: The Case of Nuclear Waste Management in Canada* (University of Toronto Press, 2008), as well as numerous scholarly articles.

Richard Kuhn is an associate professor in the Department of Geography at the University of Guelph. His research focuses on the social, political, and siting implications and strategies for nuclear fuel waste management in Canada. He has used several grants from the Social Sciences and Humanities Research Council of Canada to support this research. He has also conducted research in the Yukon, primarily on resource-management initiatives associated with First Nations' land claims. Additionally, he has recently completed a five-year, CIDA-supported project in Zhejiang Province, China, on rural development and environmental management.

Brenda Murphy is an associate professor of Geography and Contemporary Studies at the Brantford Campus of Wilfrid Laurier University, in Ontario. Her doctoral studies and subsequent SSHRC grant were focused on both the Canadian and international nuclear fuel waste context. She currently undertakes research and teaches about risk-related issues, including environmental justice, hazardous waste facility siting, emergency management, and social capital. She adopts an interdisciplinary approach focused on contexts that affect small communities and marginalized spaces, including Canada's Aboriginal peoples. Her case studies have included nuclear fuel waste management in Canada, the US, and Sweden; the e-coli water contamination crisis in Walkerton, Ontario; the Pine Lake tornado in Alberta; the Northeast Blackout of 2003 on the eastern seaboard of North America; and the effects of climate change on maple syrup production.

Anna Stanley is a lecturer in Human Geography in the Department of Geography at National University of Ireland, Galway. Dr. Stanley is interested in colonial geographies of resource management, the political economy of the nuclear fuel chain, and Aboriginal geographies of nuclear production. Her current work focuses on the political economy of knowledge production about the effects of nuclear development and radiation, and touches on themes related to imperialism, racialization, and environmental justice. She has worked with a number of First Nations and First Nation advocacy organizations as a policy analyst on issues related to experiences of nuclear industries, and is the author of a number of scholarly articles. Dr. Stanley currently teaches courses in contemporary theory, environmental justice, and risk and governmentality.

Peter Timmerman is an associate professor in the Faculty of Environmental Studies at York University in Toronto, Canada. He has been involved in environmental issues both as an academic and an activist since 1980, including as the director of the Canadian Coalition on Ecology, Ethics, and Religion (CCEER). His current concerns include climate change, nuclear waste management, spiritual traditions (particularly Buddhism) and ecology, as well as issues in environmental ethics concerning the very long-term environmental future.

Index

Printed and bound in Canada by Friesens

Set in Stone by Artegraphica Design Co. Ltd.

Copy editor: Dallas Harrison

Proofreader and indexer: Dianne Tiefensee

ENVIRONMENTAL BENEFITS STATEMENT

UBC Press saved the following resources by printing the pages of this book on chlorine free paper made with 100% post-consumer waste.

TREES	WATER	SOLID WASTE	GREENHOUSE GASES
3	**1,540**	**94**	**320**
FULLY GROWN	GALLONS	POUNDS	POUNDS

Calculations based on research by Environmental Defense and the Paper Task Force.
Manufactured at Friesens Corporation